Research for Development

The series Research for Development serves as a vehicle for the presentation and dissemination of complex research and multidisciplinary projects. The published work is dedicated to fostering a high degree of innovation and to the sophisticated demonstration of new techniques or methods.

The aim of the Research for Development series is to promote well-balanced sustainable growth. This might take the form of measurable social and economic outcomes, in addition to environmental benefits, or improved efficiency in the use of resources; it might also involve an original mix of intervention schemes.

Research for Development focuses on the following topics and disciplines:

Urban regeneration and infrastructure, Info-mobility, transport, and logistics, Environment and the land, Cultural heritage and landscape, Energy, Innovation in processes and technologies, Applications of chemistry, materials, and nanotechnologies, Material science and biotechnology solutions, Physics results and related applications and aerospace, Ongoing training and continuing education.

Fondazione Politecnico di Milano collaborates as a special co-partner in this series by suggesting themes and evaluating proposals for new volumes. Research for Development addresses researchers, advanced graduate students, and policy and decision-makers around the world in government, industry, and civil society.

THE SERIES IS INDEXED IN SCOPUS

Paola Cordera
Editor

Unfolding Made in Italy (1948-1962)

Narratives for the 21st Century

Springer

Editor
Paola Cordera
Design
Politecnico di Milano
Milan, Italy

ISSN 2198-7300 ISSN 2198-7319 (electronic)
Research for Development
ISBN 978-3-032-10117-4 ISBN 978-3-032-10118-1 (eBook)
https://doi.org/10.1007/978-3-032-10118-1

This Springer imprint is published by the registered company Springer Nature Switzerland AG
The registered company address is: Gewerbestrasse 11, 6330 Cham, Switzerland

Framing the Context: Approaches and Perspectives

In 1897, Paul Gauguin inscribed three questions in the upper left corner of one of his most ambitious canvases, *Where Do We Come From? What Are We? Where Are We Going?* Though rooted in a specific time, place, and personal vision, these words resonate beyond their Tahitian context. They pose questions of origin, identity, and purpose that continue to shape cultural and intellectual life.

This same spirit of inquiry animates the present volume: transposed into the sphere of Italian craftsmanship and design, Gauguin's questions open a space for reflection to trace the cultural and industrial lineages that have shaped Made in Italy, to explore the character and values it conveys today through archives and museums, and to imagine the trajectories it might follow in an interconnected and rapidly changing world. These questions provide an invisible framework for the following chapters, guiding the reader through history, present practices, and prospects as interwoven parts of a single narrative.

Emerging from the discussions of the international conference, "Promoting the Made in Italy Brand (1948–1960): Histories of Italian Craft and Design for the Twenty-First Century" (Politecnico di Milano, 2023)—held as part of the 2021 FARB project, VO Project. Voices of Objects. The Italian Design from Museum to Home, funded by the Department of Design at the Politecnico di Milano—this volume deliberately moves beyond the proceedings format. As Alessandro Deserti noted in the opening session, drawing on economist Gary P. Pisano's discussion of the US "industrial commons," the success of Italian design has always depended on its manufacturing base. Outsourcing and offshoring risk weakening this foundation, a concern with global resonance.

The present volume extends this debate with studies of Italian craftsmanship and design across historical and contemporary perspectives, framing them within a broader multidisciplinary field of research. Chronologically, the volume focuses on the years between the European Recovery Program (ERP) of 1948 and the apex of the "economic miracle" in 1962. This trajectory spans Italy's reconstruction, drawing on local materials for renewal, the resurgence of its industry—which expanded by 180% between the 1950s and 1960s—and its growing international reputation. The endpoint, 1962, carries symbolic weight: it was the year Dino Risi's celebrated film *Il Sorpasso* (The Overtaking) was released, with Vittorio Gassman and Jean-Louis Trintignant speeding ahead in a Lancia Aurelia B24 car, a metaphor for Italy's rapid drive into modernity. The film offered a sharp portrait of a nation suspended between carefree consumerism and shifting values, raising the question of whether this was a rupture or a continuity. Tellingly, its English title, *The Easy Life*, captured the allure and contradictions of Italy's boom years.

Engaging with this period means tracing a world in transformation, where artisans, manufacturers, artists, designers, and architects worked within dense local networks that bound production to place. As Andrea Branzi observed in his *Introduzione al design italiano* (2008), the strength of Italian design lay precisely in this fabric of relations: the capacity to merge small- and medium-scale industrial output with artisanal craft while preserving continuity with tradition.

By its very nature, postwar production invites multiple readings. Objects today are preserved and displayed across decorative arts, design, science, industrial, and corporate archives and museums, each emphasizing different facets of their identity and producing distinct narratives. This dispersal underscores the need to connect institutions and to cultivate networks among cultural organizations, researchers, and educators, as a way to sustain and renew material culture.

The volume gathers stories from diverse contexts and perspectives that reflect the ongoing search for alternative ways of preserving, representing, and experiencing "Made in Italy." They reveal the multiplicity of approaches, media, and philosophies that sustain its vitality, while also showing how Italian production continues to generate cultural value in archives, museums, cultural institutions, and creative industries. In doing so, it affirms its relevance for cultural policy and international agendas, resonating with the cultural indicators of the UNESCO 2030 Agenda for Sustainable Development.

This framework informs the ten essays collected here, which offer a transdisciplinary reading of the Italian design system through the lenses of architecture, cinema, decorative arts, craft, furniture, product, and interior design. Rather than aiming at a totalizing history, they present intersecting perspectives that illuminate how diverse approaches and media have shaped its cultural identity. Central to this discussion is heritage: how material traditions and places can be preserved, reimagined, and transmitted in the digital age.

Whether examining the past or engaging with present practices, the contributions demonstrate how multiple trajectories converge in a shared story. That story underscores the deep ties between craftsmanship and community, while also addressing new ways of reinterpreting and transmitting heritage across local and global scales, and across physical, digital, and virtual spaces.

The essays are organized in three interconnected thematic parts. The first, "On the Uniqueness of Italian Design," adopts a historical perspective and examines the different contexts through which the image of Made in Italy was shaped in the postwar decades. Exhibitions abroad acted as instruments of cultural diplomacy, newsreels brought design into the everyday imagination of mass audiences, and hotels within Italy offered visitors—both domestic and foreign—tangible encounters of an emerging Italian lifestyle. Taken together, these examples show that Made in Italy was never confined to single objects but was constructed through a mosaic of practices, images, and spaces that intertwined modernity and tradition, leaving a legacy that still informs cultural memory today.

Opening this section, Paola Cordera examines how economic and cultural promotion intersected in shaping the international recognition of Made in Italy after World War II. Beyond foreign aid and industrial reorganization, Italy's recovery relied on projecting a distinctive image abroad, pursued through exhibitions, art galleries engaged in cultural brokerage, department store collaborations, and promotional campaigns. These coordinated efforts turned products into cultural ambassadors, presenting art, design, and craftsmanship as facets of the same identity. The interplay of institutions, entrepreneurs, and designers transformed fragmentation into a resource, enabling Italy to establish itself as both a commercial force and a cultural presence. By the 1960s, Italian design had become a strategy for export growth and a narrative of national identity, a foundation on which later representations of Made in Italy would build.

In a complementary way, Giampiero Bosoni and Marta Elisa Cecchi examine how visual storytelling—through television, documentaries, and above all *La Settimana Incom* (The Incom Weekly)—shaped the image of Italian craftsmanship, design, and culture on both national and international stages. These media did more than display elegant and innovative products: they wove them into Italy's postwar identity, turning Made in Italy into a symbol of quality, sophistication, and modern lifestyle. The essay traces how newsreels and broadcasts connected design to collective imagination and how they shaped domestic perceptions, embedding design into everyday cultural imagination.

Extending this perspective from media to real space, Francesco Scullica and Elena Elgani explore how Italian hotels, from the 1950s to the present, have acted as catalysts for promoting Made in Italy design and lifestyle. Conceived in the postwar decades by leading architects and designers, they functioned as living showrooms of creativity, merging interiors, furnishings, and services into a modern vision of leisure and tourism. Central to this story is celebrated architect Gio Ponti, who, through projects such as the Parco dei Principi Hotel in Sorrento, reimagined hospitality as a site of experimentation and cultural expression. The essay also considers how these hotels operate today as laboratories of design, where the legacy of postwar creativity intersects with contemporary notions of Italian style and international tourism.

Building on this interplay between legacy and the present, the second part, "Archives for Memory and for the Future," explores how archives safeguard and reframe Made in Italy narratives in physical and digital forms. They emerge as dynamic repositories that protect memory and activate processes of interpretation and participation, strengthening ties among institutions, researchers, and educators in constructing cultural identity. Traditional instruments and innovative channels—from curated exhibitions to digital platforms and social media—serve as tools that extend access and foster transdisciplinary engagement, keeping heritage alive through renewed forms of use and meaning. At the same time, they address the complexities of a heritage that extends beyond canonical masterpieces to include replicas, copies, and objects of serial production, whose value often lies in their relational and contextual significance. Making such materials visible may permit connections with resources preserved in other institutions, opening the way to new readings and previously unseen patterns of meaning.

Within this framework, Maria Teresa Feraboli examines how Milan's CASVA (Centro di Alti Studi sulle Arti Visive) has developed innovative strategies to communicate and valorize its design archives. Since its foundation, the center has progressively expanded its holdings, incorporating collections of figures such as Luciano Baldessari, Roberto Sambonet, and Nanda Vigo. It recently relocated to its new headquarters at QT8, a district emblematic of Milanese modernity. Alongside its traditional mission of research and preservation, CASVA has embraced social media, employing platforms such as Instagram and Facebook to reinterpret archival materials for contemporary audiences. Through curated online exhibitions, thematic narratives, and interactive features, it demonstrates how cultural heritage can be resemanticized and made accessible to diverse publics, renewing and projecting Made in Italy's legacy internationally.

Rosa Chiesa and Teresita Scalco, in turn, focus on the Archivio Progetti at the Università Iuav di Venezia as a research and archival center of primary importance for studying Italian design and craftsmanship. Through collections such as those of Paolo De Poli, Giorgio Casali, and Luca Meda, the archive provides an invaluable perspective on the evolution of Made in Italy, spanning from artistic experimentation to industrial production and photographic documentation. As the authors underline, the institution preserves these legacies and reinterprets them through research initiatives, exhibitions, and digital projects—including ApOnline and the Petit Tour platform—that broaden access and reframe Italian heritage for contemporary audiences. In this way, the Archivio Progetti emerges as a dynamic laboratory for renewing and disseminating design studies in the digital era.

Finally, Sandra Costa and Irene Di Pietro turn to Cirulli Foundation, established in Bologna in 2015 after decades of acquisitions dedicated to Italian visual culture from the early twentieth century to the economic boom. They highlight how the Foundation's holdings—one of the most important private collections of postwar Italian graphic art and design—are framed through a distinctive museographic and museological vision. Their discussion addresses the building designed by Achille and Pier Giacomo Castiglioni, the curatorial strategy of "capsule collections" developed by Jeffrey Schnapp as an "animated archive," as well as the interplay between the permanent collection and temporary exhibitions, conceived through a narrative method that combines storytelling, visual cues, and scholarly research.

Drawing on the previous focus on archives and memory, the third part, "Made in Italy on Display Inside and Outside Museums," shifts attention from conservation to presentation. Here, design is considered a heritage to be safeguarded, moving beyond the traditional view of design as merely the memory of objects and a cultural resource that takes shape through exhibitions, installations, and digital initiatives. The contributions assembled in this part highlight how curatorial strategies, display practices, and innovative technologies transform heritage into participatory experiences, reinforcing the global resonance of Italian creativity. Some of these projects—often supported by publicly funded programs—further underscore the importance of collaboration between academic research and archival institutions in redefining the modes of mediation and cultural valorization, shifting from the concept of a visit to one of an experience, where audience participation and inclusion are crucial.

Gianluca Carella, Francesco Zurlo, and Carlo Martino examine the exhibition *Identitalia. The Iconic Italian Brands*, presenting over one hundred companies as emblems of national identity. Through the interplay of archival sources, product histories, and corporate narratives, they investigate how industrial heritage can be mobilized to construct shared memory. Their analysis highlights the curatorial methodology, which combined digital ethnography with direct engagement with firms, revealing heterogeneous archival practices that shaped the exhibition and posed challenges of documentation, accessibility, and brand self-representation. By turning such constraints into narrative opportunities, *Identitalia* emerges as a case study of how design exhibitions can mediate between past and present, renewing the discourse on national identity in contemporary Italy.

While *Identitalia* foregrounds the role of corporate heritage in shaping collective memory, Ico Migliore shifts the attention to the transversality of Italian design, conceived as a process of encounter and contamination between disciplines, cultures, and contexts. At the center of his essay is the traveling exhibition *Coats! Max Mara* which transformed the company's archive and heritage into a narrative space of dialogue between industrial production, art, design, and communication. The project underscores how, far from being a passive repository, the archive can catalyze innovation and cultural exchange, as memory and creativity converge in immersive and polyphonic experiences. More broadly, the essay situates exhibition design as a privileged field of experimentation, where the dialogue between design culture and industrial culture reveals design itself as a continuous act of negotiation capable of generating new forms of identity and imagination.

If Migliore highlights the experiential power of exhibitions, Mauro Ceconello and Davide Spallazzo turn to the potential of digital tools in reactivating design heritage. Their contribution investigates how dormant archives and former studios can be revitalized through strategies that connect material traces with contemporary audiences. Within this framework, they present two case studies: the digitisation of Giovanni Sacchi's archive, which reconstructs design trajectories by linking sketches, drawings, photographs, and prototypes, and Looking for Achille Castiglioni, a location-based mobile application that guides visitors through a geolocated narrative of Castiglioni's works in Milan. Together, these projects illustrate how digital technologies can transform archival sources into accessible and dynamic cultural experiences, enabling new forms of interpretation and participation in the legacy of Italian design.

Ilaria Bollati and Luisa Collina turn to a less visible but equally vital aspect of heritage: the intangible dimension of Italian design, rooted in relationships, oral traditions, and informal spaces of encounter. Their essay examines how cafés, clubs, and everyday meeting places in postwar Milan nurtured creativity beyond the boundaries of studios and institutions. This fragile legacy, resistant to conventional archival practices, is sustained through oral transmission and digital infrastructures conceived as sensitive preservation tools. By integrating mapping, podcasts, and sound books, the project *Milano In-formale* creates an original constellation of voices and memories that reveal design as a collective process embedded in urban life. In this way, intangible heritage emerges not as nostalgia, but as a living resource that continues to renew design practices and cultural narratives today.

In reviewing the central themes of this volume, it becomes clear that Made in Italy thrives at the crossroads of tradition and innovation, standing as a significant cultural asset. The essays emphasize the importance of viewing postwar Italian design systematically, echoing art historian Andrea Emiliani's 1985 concept of a "widespread museum" in reference to Italy's historical heritage. They show how institutions adapt to contemporary demands by integrating digital tools, media platforms, new technologies, and alternative narratives. Such developments resonate with ICOM's (International Council of Museums) 2022 definition of museums as accessible, inclusive spaces that serve society.

Rather than composing a conventional story, the three sections map a series of crossings where Italian production is redefined: between national image and international circulation, between the permanence of museums and archives and the ephemerality of media, and between material artifacts and intangible practices. What holds these contributions together is a shared attention to mediation—exhibitions, films, displays, digital tools, oral testimonies—understood not as neutral channels but as formative devices that shape what can be remembered, valued, and projected. The volume thus advances a critical reflection on how Made in Italy has been constructed and reconstructed over time and how its legacy can be mobilized in the present. More than a story of objects—or a "system of objects," to use the expression of sociologist and philosopher Jean Baudrillard—it proposes a genealogy of practices and narratives that interrogates the cultural, political, and economic stakes of "the Italian way," while opening new perspectives on its global circulation.

The book points to promising directions for further research, particularly by raising awareness of the relevance of archives and museums as custodians of memory, preservation, and public access. The emphasis on quality and durability that characterized postwar Italian design—rooted in local production, materials, territories, and cultural traditions—and documented in these institutions offers valuable models for contemporary practice. Collaboration between cultural institutions, scholars, professionals, industries, and stakeholders will sustain this vitality and reinforce Made in Italy as a cultural heritage and economic resource.

Far from being confined to history, Made in Italy embodies a dynamic tradition that continues to shape the present and future of design, offering innovative solutions yet rooted and culturally resonant on the international stage. It lives not only in objects but in the archives, museums, places, and practices that preserve and reimagine its legacy.

Politecnico di Milano
Milan, Italy

Paola Cordera

Contents

On the Uniqueness of Italian Design

Instructions: Ensure "Made in Italy" Is Stamped on Every Single Item ... 3
Paola Cordera

Made in Italy Written in the Stars (and Stripes): Italian Craft's Global Influence in Vintage Newsreels and Documentaries ... 17
Giampiero Bosoni and Marta Elisa Cecchi

Icons of Memory and Modernity: Italian Hotels of the 1950s and the Promotion of Italian Lifestyle ... 31
Francesco Scullica and Elena Elgani

Archives for the Memory and for the Future

Design Archives, Made in Italy, and Communication in CASVA's Recent Activities ... 49
Maria Teresa Feraboli

Valorization, Promotion and Communication of Made in Italy in the Archivio Progetti, Università Iuav di Venezia ... 61
Rosa Chiesa and Teresita Scalco

The Tale of Made in Italy and the Massimo and Sonia Cirulli Foundation: The Relationship Between Collections and Exhibition Choices ... 73
Sandra Costa and Irene Di Pietro

Made in Italy on Display Inside and Outside Museums

***Identitalia. The Iconic Italian Brands*: Notes on Exhibition Challenges** ... 89
Gianluca Carella, Francesco Zurlo, and Carlo Martino

Archives and Project: The Case of the Traveling Exhibition *Coats! Max Mara* 103
Ico Migliore

Digital Technologies to Promote and Valorise Design Culture 115
Mauro Ceconello and Davide Spallazzo

Milano and the Everyday Informal: Mapping the Intangible Heritage of Italian Design 129
Ilaria Bollati and Luisa Collina

Contributors

Ilaria Bollati Politecnico di Milano, Milan, Italy

Giampiero Bosoni Politecnico di Milano, Milan, Italy

Gianluca Carella Politecnico di Milano, Milan, Italy
MIT Morningside Academy for Design-Massachusetts Institute of Technology, Cambridge, MA, USA

Marta Elisa Cecchi Politecnico di Milano, Milan, Italy

Mauro Ceconello Politecnico di Milano, Milan, Italy

Rosa Chiesa Università Iuav di Venezia, Venice, Italy

Luisa Collina Politecnico di Milano, Milan, Italy

Paola Cordera Politecnico di Milano, Milan, Italy

Sandra Costa Alma Mater Studiorum, Università di Bologna, Bologna, Italy

Irene Di Pietro Alma Mater Studiorum, Università di Bologna, Bologna, Italy

Elena Elgani Politecnico di Milano, Milan, Italy

Maria Teresa Feraboli Politecnico di Milano, Milan, Italy

Carlo Martino Sapienza Università di Roma, Rome, Italy

Ico Migliore Politecnico di Milano, Milan, Italy

Teresita Scalco Università Iuav di Venezia, Venice, Italy

Francesco Scullica Politecnico di Milano, Milan, Italy

Davide Spallazzo Politecnico di Milano, Milan, Italy

Francesco Zurlo Politecnico di Milano, Milan, Italy

List of Figures

Instructions: Ensure "Made in Italy" Is Stamped on Every Single Item

Fig. 1 S.E.T. Apollon-Rome (printer), *Verità sul piano Marshall* (The Truth behind the Marshall Plan), 1950. Milan, Private collection. 5

Fig. 2 United Press-ACME, An ox-drawn cart meets a speeding motorcycle, symbolizing the clash between rural tradition and modern technology in the Siena countryside, 1952. Milan, Private collection. 8

Fig. 3 A.C. Cecchini, Chairman of an exhibition's finance committee, and A.E. Scarpelli, Co-chairman, examine a Sicilian donkey cart for carrying vegetables, painted with scenes of Columbus's discovery of America by Danea of Palermo, 1951. Milan, Private collection. 10

Fig. 4 Keystone Press Agency, Clare Boothe Luce, US Ambassador to Italy, at the Florence Artisan Fair, where an exhibitor presents her with a straw bonnet, 1955. Milan, Private collection 13

Made in Italy Written in the Stars (and Stripes): Italian Craft's Global Influence in Vintage Newsreels and Documentaries

Fig. 1 Header of the La Settimana Incom newsreel, *Due civiltà s'incontrano*, 1957. Retrieved from Istituto Luce Archive, Rome. 22

Fig. 2 Giulio Minoletti, Front of the lead carriage of the ETR 300 train, known as Settebello [One's best card], produced by Breda Ferroviaria, 1952 23

Fig. 3 Olivetti shop in New York, 584 Fifth Avenue, designed by BBPR in 1954, in the Incom newsreel, *Due civiltà*

s'incontrano, 1957. Retrieved from Istituto Luce Archive, Rome 27
Fig. 4 Gio Ponti, espresso coffee machine, known as La Cornuta [Horned], produced by La Pavoni, 1949. 28

Icons of Memory and Modernity: Italian Hotels of the 1950s and the Promotion of Italian Lifestyle

Fig. 1 The outdoor swimming pool with an organic shape. (Image courtesy of Hotel Parco dei Principi Sorrento). 40
Figs. 2 and 3 (above) and (below) In common living spaces on the ground floor, the patterns and the materials expressions of horizontal and vertical surfaces (decorated tiles, majolica elements, and ceramic pebbles) are combined. (Image courtesy of Hotel Parco dei Principi Sorrento) 41
Fig. 4 All the hotel rooms are characterised by personalized floor decoration, equipped bed headboard and selected furniture. (Image courtesy of Hotel Parco dei Principi Sorrento). 42

Design Archives, Made in Italy, and Communication in CASVA's Recent Activities

Fig. 1 Poster of the exhibition *Milano Città Immaginata* (2015) on the Facebook CASVA Homepage 52
Fig. 2 From the Facebook album *Compasso d'Oro*: the fish kettle (1957), awarded in 1970. 54
Fig. 3 From the monographic album dedicated to designers: *Augusto Magnaghi and Mario Terzaghi and the design of the '50s*: the Bilux sofa, Busnelli, c.1955. Comune di Milano—CASVA ©Photograph by Aldo Ballo 56
Fig. 4 The post reads: "This August, a month of holidays and travel, we will take you with us to discover some projects kept in our archives and made in foreign countries or that show an international scope". The 1954 exhibition dedicated to Japan by la Rinascente, when Amneris Lis was the art director 58

Valorization, Promotion and Communication of Made in Italy in the Archivio Progetti, Università Iuav di Venezia

Fig. 1 The historic headquarters of the Archivio Progetti in Venice, featuring Massimo Scolari's iconic "Wings". Università Iuav di Venezia/Umberto Ferro. 62
Fig. 2 Exhibition setup of the education display *De Poli. Artigiano e designer*, curated by Alberto Bassi, Valeria Cafà, and

Teresita Scalco, Archivio Progetti Iuav, Venice, May–July 2014. Università Iuav di Venezia/AP 65
Figs. 3 and 4 Images from the exhibition *Luca Meda. La felicità del progetto*, curated by Nicola Braghieri, Rosa Chiesa, Serena Maffioletti and Sofia Meda, Triennale, Milan, May–June 2013. (Courtesy of Armando Bernacchi) 67
Fig. 5 Petit Tour digital platform. 68
Fig. 6 Petit Tour edited by Anna Giambellini and Cecilia Marcolini created during the master's course *Metodologie della ricerca: fonti, archivi e reti* by Rosa Chiesa in 2022–2023 69

The Tale of Made in Italy and the Massimo and Sonia Cirulli Foundation: The Relationship Between Collections and Exhibition Choices

Fig. 1 The Massimo and Sonia Cirulli Foundation's front door with Walter Gropius's quote from the preface of the Italian edition of *Scope of Total Architecture*. 78
Fig. 2 Massimo and Sonia Cirulli Foundation. Capsule AA14MW, dedicated to the *Monotypes of Venturino Venturi* 80
Fig. 3 Massimo and Sonia Cirulli Foundation. Second floor of Achille Castiglioni's building 81
Fig. 4 Massimo and Sonia Cirulli Foundation. Capsule dedicated to *Fiorucci: pop revolution* 82

***Identitalia. The Iconic Italian Brands*: Notes on Exhibition Challenges**

Fig. 1 Overview of the exhibition layout. Rome, Palazzo Piacentini, 2024 93
Fig. 2 View of the different areas divided according to the adopted phases of the day. Rome, Palazzo Piacentini, 2024 94
Fig. 3 Overview of the exhibition area dedicated to the theme of "Night." Rome, Palazzo Piacentini, 2024. 95
Fig. 4 Overview of the typologies of artefacts used in the exhibition area dedicated to the theme of "Night." Rome, Palazzo Piacentini, 2024 95

Archives and Project: The Case of the Traveling Exhibition *Coats! Max Mara*

Fig. 1 The dome. *Coats! Max Mara* Exhibition. Multimedia and Graphic Design by Migliore+Servetto, Dongdaemun Design Plaza (DDP). Seoul, November 2017. (Photo by Jae Young Park) 104

Fig. 2 *The Creative Studio. The Democratic Coat, Designers, and Pop* (1960s), *Coats! Max Mara*, Exhibition, Multimedia and Graphic Design by Migliore+Servetto, Dongdaemun Design Plaza (DDP). Seoul, November 2017. (Photo by Migliore+Servetto) . . . 107
Fig. 3 *The Creative Studio. The Democratic Coat, Designers, and Pop* (1960s), *Coats! Max Mara* Exhibition, Multimedia and Graphic Design by Migliore+Servetto, Dongdaemun Design Plaza (DDP). Seoul, November 2017. (Photo by Migliore+Servetto) . . . 108
Fig. 4 Interiors of the Dome, with the digital installation of Yiyun Kang. *Coats! Max Mara* Exhibition, Multimedia and Graphic Design by Migliore+Servetto, Dongdaemun Design Plaza (DDP). Seoul, November 2017. (Photo by Jae Young Park) . . . 110

Digital Technologies to Promote and Valorise Design Culture

Fig. 1 Study sketches for the Gibigiana Lamp, digitised with Kodak target. (Image courtesy of Fondazione Achille Castiglioni—SIRBEC) . . . 121
Fig. 2 Sketches and brainstorming on the Gibigiana lamp project. (Image courtesy of Fondazione Achille Castiglioni—SIRBEC) . . . 124
Fig. 3 LfAC Looking for Achille Castiglioni Mobile app. . . . 125

Milano and the Everyday Informal: Mapping the Intangible Heritage of Italian Design

Fig. 1 Conceptual collage, visualizing Milan's urban fabric's subtle, unplanned design origins. . . . 132
Fig. 2 Initial visualization tests of possible key sites explored during the research. (Images by Atelier Fyumi). We explore key visual language elements to align the illustrations with all project outputs . . . 134
Fig. 3 Conceptual collages showing the interplay of anecdote, oral storytelling, podcast, and mapping . . . 137

On the Uniqueness of Italian Design

Instructions: Ensure “Made in Italy” Is Stamped on Every Single Item

Paola Cordera

Abstract This essay examines the role of cultural and promotional initiatives in consolidating the international reputation of Italian production in the postwar decades. From the late 1940s, national reconstruction depended on financial aid, industrial reorganization, and its ability to access foreign markets, particularly the United States. Customs regulations created obstacles, yet simultaneously contributed to affirming the Made in Italy brand as a recognizable mark of identity in the subsequent years. Through exhibitions in museums, department store events, and promotional programs across the US (and later in Europe), Italian goods were presented as both commodities and cultural artifacts, linking modern production with traditions of craftsmanship. Networks of institutions, entrepreneurs, artisans, artists, architects, critics, and designers fostered collaborations that transformed structural weaknesses into opportunities, shaping the Made in Italy brand as a symbolic and commercial presence abroad. By the 1960s, building on earlier efforts, Italian production had matured into both a commercial strategy and a cultural narrative, establishing a legacy that continues to shape the present.

Keywords Exhibitions · Collaborative networks · Made in Italy · Postwar production · Art galleries

1 From Recovery to International Recognition

In the late 1940s, as Italy confronted the immense task of reconstructing its political, economic, and cultural life in the aftermath of war, opening national production to foreign markets rapidly emerged as a central priority.

Public institutions and private initiatives were engaged in rebuilding the industrial sector, severely weakened by shortages, infrastructural collapse, and

P. Cordera (✉)
Politecnico di Milano, Milan, Italy
e-mail: paola.cordera@polimi.it

P. Cordera (ed.), *Unfolding Made in Italy (1948-1962)*, Research for Development, https://doi.org/10.1007/978-3-032-10118-1_1

fragmentation. Within this context, measures such as the Eximbank Loan (1947) and the European Recovery Program (ERP, 1948–1952)—better known as the Marshall Plan—proved crucial to sustaining recovery by providing financial assistance, raw materials, and technical expertise, thereby linking reconstruction with the drive toward modernization (Fig. 1). At the same time, a range of initiatives contributed to Italy's political alignment within the Western sphere of growth, consistent with the broader dynamics of the Cold War. From this twofold perspective, the development of exports was indispensable, fostering recovery while positioning Italy as an active player in the renewed international order.

The United States stood out as the most attractive market among all potential outlets: its industrial and commercial system had been undiminished by the war, and its consumer base remained intact. In 1949, Clemente Boniver, economic counselor at the Italian Embassy in Washington, emphasized to his counterparts in Italy the obstacles hindering the establishment of the Italian industry in the US market and the measures necessary to address them (Beltramini 1949a). He stressed that taking advantage of the postwar recovery required market expansion, effective communication with American consumers, streamlined logistics (from packaging to transportation), and consortia designed to ease the burdens of US customs procedures.

In doing so, Boniver identified the fragmentation of Italian production and export structures as the central issue. He argued that the real challenge lay in coordinating the efforts of entrepreneurs, trade associations, and public institutions, so that what remained unattainable for individual firms could be achieved through a collective organization capable of acting on behalf of Italian companies. Similar concerns were articulated concurrently in the *Monthly Review* of the American Chamber of Commerce for Italy, which drew attention to structural problems often neglected by exporters: the international protection and registration of the *marchio di fabbrica* (trademark), limited knowledge of the US market, lost orders, fragile shipping and delivery systems, poor inspection and packaging practices, and the absence of stable pricing systems (Beltramini 1949b).

One notable issue was the American requirement—applying to all imports—that every item be clearly marked with its country of origin. The label "Made in Italy" had to be permanently affixed to every product, whether engraved in wood, stamped into leather, or written beneath the ceramic glaze, in such a way that it could not be removed. Like a mantra, instructions such as "Remember to stamp 'Made in Italy' on every single item" appeared at the bottom of purchase orders: an indication that these practices were still far from being integrated into routine production processes. For many items, particularly handcrafted goods, delays were caused by reworking requests: products had to be returned to manufacturers, withheld from sale, or accepted with an additional surcharge of around 10% due to improper labeling.

Such complications significantly constrained Italian exports to the American market. Although the regulation was intended to ensure transparency, it posed significant challenges to Italian firms, as compliance with the prescribed standards demanded resources and organizational capacities that exceeded the means of many producers. Yet, in a paradoxical turn, this requirement fostered the consolidation of "Made in Italy" as a recognizable and enduring brand in the international market, even if the process unfolded slowly and unevenly.

Fig. 1 S.E.T. Apollon-Rome (printer), *Verità sul piano Marshall* (The Truth behind the Marshall Plan), 1950. Milan, Private collection

Fig. 1.1 (continued)

2 Selling with Art(istry)

Exhibitions and trade fairs emerged as crucial cultural diplomacy instruments and vehicles for projecting Italian products into international markets, where they acquired visibility through carefully staged presentations to American consumers. In 1950, the international fairs of Chicago and Detroit marked a turning point by admitting foreign exhibitors alongside American firms, thereby creating new opportunities for transatlantic exchange among administrations, industries, and buyers. The vast geography of the United States further favored the model of the traveling exhibition, a format pioneered by France with *France Comes to You* (1948–1951), which advanced commercial aims through events in hotels across the country.

However, Italy pursued a different strategy. Rather than presenting its goods as mere commodities, they were embedded in a celebratory narrative that linked modern production to ancestral traditions, valorized as authentic and virtuous. This narrative also amounted to an implicit renunciation of the recent Fascist past, which was cast as a rupture to be overcome: the ambition was to demonstrate that new production was not only modern and competitive, but also the continuation of centuries of creativity and craftsmanship (Cordera and Turrini 2025). This ambivalence was vividly captured by contemporary photography, where many images staged the tension between old and new, at once bearing witness to the country's contradictions and, at times, celebrating them.[1] Such perspectives emerge, for example, in the work of Curtis Bill Pepper, David Seymour, or photographers working for the United Press Association (Fig. 2).

The same narrative was subsequently articulated through institutional platforms abroad. Inaugurated in April 1947, the House of Italian Handicraft (HIH) in New York emerged as a pivotal venue for promoting Italian production in the United States. Established by Handicraft Development Inc. (HDI) and CADMA (Commissione Assistenza Distribuzione Materiali Artigianato, or Commission for Assistance and Distribution of Handicraft Materials), it offered American buyers a curated entry point into Italian craftsmanship, positioning it within an international marketplace. The HIH Board of Directors composition was strategically calibrated to maximize the visibility and promotion of Italian goods in the American market: its members embodied a convergence of expertise in retail, publishing, cultural mediation, and transatlantic trade. Chaired by David Freudenthal—former vice-president and treasurer of Bloomingdale Bros. and assistant to the head of the ECA (Economic Cooperation Administration) mission in Rome—the board included Adam Gimbel of the Gimbels chain, Dorothy Shaver—president of Lord & Taylor—, Roger William Straus Jr.—cofounder of Farrar, Straus and Giroux publishing company—, and Ercole Sozzi, owner of Fontana Hollywood Wine Importers and director of the American Chamber of Commerce for Trade with Italy. This distinguished network linked the HIH to primary American retail circuits, channels of

[1] For a comparable representation in the *Settimana Incom* newsreel, see the essay by Bosoni and Cecchi in this volume.

Fig. 2 United Press-ACME, An ox-drawn cart meets a speeding motorcycle, symbolizing the clash between rural tradition and modern technology in the Siena countryside, 1952. Milan, Private collection

cultural mediation, and Italian business interests, thereby ensuring a powerful nexus of commerce, culture, and diplomacy. The strategy found concrete expression in the HIH's temporary exhibitions, which showcased a broad spectrum of Italian products—from ceramics to furniture and glassware—framing Italian craftsmanship at once as a marketable commodity and a symbolic resource charged with aspirational appeal (Cordera 2022).

If the HIH laid the institutional foundations, the traveling exhibition *Italy at Work: Her Renaissance in Design Today* (1950–1953) marked a new phase, consolidating Italy's position in the international marketplace. Organized under the auspices of the CNA (Compagnia Nazionale Artigiana, or National Artisans Company), it introduced the American public to a curated selection of Italian products that combined traditional and modern craftsmanship and design, emphasizing continuity with Italy's cultural heritage. One of the exhibition's most compelling features was the display of "five special interiors", designed by architects Luigi Cosenza, Roberto Menghi, Carlo Mollino, and Gio Ponti, together with artist and stage designer Fabrizio Clerici. These fully furnished environments were conceived not merely as demonstrations of design but as embodiments of a lifestyle, offering a vision of Italian domestic culture that invited American audiences to imagine routines enriched by beauty and artisanal skill.

A broad spectrum of Italian craftsmanship was represented, from renowned brands such as Salvatore Ferragamo—the "Shoemaker to the Stars"—, the Richard Ginori Ceramic Company, and the Venini Glass Company, to smaller, emerging firms. This diversity underscored Italy's capacity to combine artistry with industry, projecting an image of a production system ranging from artisanal workshops to larger-scale manufacturers.

As Cervelli from the American Chamber of Commerce observed: "The presentation of Italian handicraft to American consumers through [...] exhibitions [...] have been of enormous value in bringing to the attention of the American public the skill and creative genius of the Italian craftsman; and at the same time bringing to the Italian industry the realization of the great potential of the United States market and the necessity of marking goods to suit its tastes in design and utility. There is no doubt that Italian handicraft now has a firmly established position in the American market, and, in fact, has set buying standards for many types of these goods. The future, therefore, should bring a steady and progressive increase in sales" (Cervelli 1952, p. 34). His assessment underscored that exhibitions such as *Italy at Work* functioned as cultural showcases and strategic instruments of market penetration, positioning Italian production as both functional and highly desirable in the United States.

Products first displayed in cultural settings soon reappeared as commodities in department stores and shopping malls, often framed within promotional "Italian events" aimed at a broader consumer audience. In this way, cultural display became directly entangled with commercial promotion, anticipating later forms of lifestyle marketing. Prominent department stores—including Macy's,[2] Kaufmann's, Lord & Taylor, Abraham & Straus, Marshall Field, McCurdy's, Jackson's, Meier & Frank, Boyd's, and Famous-Barr Co.—capitalized on the publicity surrounding the museum exhibitions to promote Italian goods, reinforcing the nexus between cultural identity and commerce. Sales campaigns were sometimes accompanied by live demonstrations, further sustaining consumer interest in the *saper ben fare* (mastery of craftsmanship) adapted to the needs of a modern economy. In doing so, they paralleled the immersive "atmosphere rooms" devised by art historian and director Alexander Dorner for museums, offering audiences experiences that blurred the boundaries between cultural display and commercial marketing.

Notably, the typology of objects promoted through the different exhibitions corresponded to a list compiled by the National Retail Dry Goods Association and the Association of Buying, which identified the products considered particularly suitable—or even preferred—for the US market (Beltramini 1950). This comprehensive inventory included textiles (silk, wool, cotton, rayon, hemp, and linen), knitwear and hosiery, lace and embroidery, gloves, hats, umbrellas, leather goods, footwear, gold and silver jewelry, ceramics, porcelain and terracotta, marble and alabaster items,

[2] Italian firms wishing to participate in the exhibition were required to submit proposals to Macy's Italian branch in Florence. The American department store committed to sending 19 purchasing agents to Italy to orient exhibitors toward its commercial requirements and to encourage the development of products adapted to the American market.

Fig. 3 A.C. Cecchini, Chairman of an exhibition's finance committee, and A.E. Scarpelli, Co-chairman, examine a Sicilian donkey cart for carrying vegetables, painted with scenes of Columbus's discovery of America by Danea of Palermo, 1951. Milan, Private collection

lighting fixtures, wooden products, musical instruments, cork articles, phonograph records, dolls and toys, sporting goods, bicycles, motor scooters, sewing machines, typewriters, optical and photographic equipment, as well as prints and gift items.

Although the list included modern devices and despite efforts to foreground Italy's modern productive image, contemporary newspaper accounts nonetheless revealed a marked interest in the more folkloric artifacts of Italian production (Fig. 3).[3]

Such items were also privileged by American intermediaries and buyers not only because of their appeal, but also because they posed no challenge to US industrial output. This selective gaze—aptly described as a "colonizing eye"—foregrounded artifacts that bore not the imprint of the "machine" but rather that of "humanity," of the "hand" bestowed upon Italians by God and admired, even envied, abroad (Calzini 1948). This reception mode reveals the ambivalence of Italy's image: at once modernizing yet persistently framed through an exoticizing lens. Such an interpretive framework reinforced the allure of Italian goods while simultaneously safeguarding American markets.

[3] Significantly, Raffaele Bedarida has shown how Lucio Fontana's public image shifted across contexts. In fact, in Europe this artist engaged performatively with national identity, while in the United States he played into the clichés of *Italianità*, or "Italianness". This contrast was epitomized by the 1950 issue of *Interiors* magazine, which featured a relaxed, sun-tanned Fontana at work in Albisola (Savona) rather than the formal figure familiar from Spatialist publications (Bedarida 2023).

Yet promotional events did more than sustain traditional charm: they opened new possibilities beyond official or pre-established channels. Among them were collaborations between Italian architects—such as Gio Ponti, Carlo Mollino, and Ico Parisi—and American firms like Altamira and Singer & Sons, which developed new furniture lines expressly for US consumers (Casali 2020; Altea 2024). Here, the traditional image intersected with a parallel narrative of modernist innovation, allowing Italian design to be reframed for a mass American market.

This promotional drive expanded through a series of initiatives that carried Italian production to broader audiences. For instance, *Italy Today: New York Fair of Italian Manufacturers* (1952) targeted buyers in the nation's commercial capital, while the *Modern Movement in Italy: Architecture and Design* exhibition (1953) underscored the country's influential role in shaping contemporary architecture and design (Casali 2022). Shortly thereafter, the Smithsonian's *Italian Arts and Crafts Show* (1955–1957) extended this reach by touring not only major cities—like Chicago or San Francisco—but also smaller centers across the Midwest and South, regions that had not previously hosted large-scale Italian exhibitions. By moving beyond traditional cultural hubs, such initiatives reinforced Italian production's visibility, embedding it within elite and popular consumer landscapes. Taken together, they laid the groundwork for the institutionalization of Made in Italy as both a cultural and commercial brand in the United States.

Moreover, they consolidated Italy's international reputation, affirming its reliability as a political and economic partner while promoting a renewed cultural identity abroad. Parallel promotional efforts unfolded across Europe and beyond. Notable examples include the *Mostra dell'artigianato italiano* (Cairo 1951 and Helsinki 1951), *Art décoratif italien* (Paris 1952), *Moderne Italienische Keramik* (Munich, Stuttgart, Karlsruhe, and Düsseldorf 1953), *Nutida italiensk konst* (Stockholm and Helsinki 1953), *Forme nuove in Italia: italienischer Stil vom Kunsthandwerk zur Industrie* (Zürich and Düsseldorf 1954-1955), *Selected examples of Italian Industrial Design* (London 1955), and *Formes et idées d'Italie* (Paris 1957). Often organized in collaboration with institutions such as the CNA or the Biennale di Venezia, these exhibitions extended Italy's cultural diplomacy beyond the Atlantic, adapting display strategies to diverse contexts. They repositioned Italian identity by negotiating the transition from traditional craft to modern aesthetics, aligning their narratives with host countries' cultural expectations and commercial priorities.

The strategies tested in these exhibitions also echoed a broader international debate on display practices. In fact, in those same years, in its journal *Museum*, the International Council of Museums (ICOM) discussed how display practices might reach wider audiences. Particular attention was given to strategies that opened museums onto the street with outward-facing displays designed to attract passers-by and draw them inside. Collaboration with shops was also highlighted to extend the museum's visibility and promote temporary exhibitions.[4]

[4]Alongside examples from the United States, the journal pointed to comparable initiatives in France, Spain, and Norway.

3 Hope Lingered in the Air, Tangible as Stone

Italian exhibitions abroad played a decisive role in the nation's postwar rehabilitation. Backed by central state administration, such events drew on the support of a wide range of actors, including Italian cultural institutes, Chambers of Commerce, ICE (Istituto per il Commercio Estero or Institute for Foreign Trade), ENAPI (Ente Nazionale per l'Artigianato e le Piccole Industrie, or National Agency for Handicraft and Small Industries), APE (Artigianato Produzione Esportazione, or Handicraft Production Export), CIM (Centro Italiano della Moda, or Italian Fashion Center), CADMA, and the CNA. Cultural institutions such as the Milan Triennale and the Venice Biennale were likewise called upon to curate and stage international showcases.

Although these organizations represented diverse agendas, they converged around shared political, economic, and cultural priorities. Their involvement eased the cumbersome organizational process, streamlined logistics, and enabled coordinated strategies, while also reinforcing connections among entrepreneurs, artisans, manufacturers, artists, designers, and architects. These alliances were emblematic of the Italian model of production, deeply embedded in the cultural and territorial fabric of the country. As Vignozzi Paszkowski has recently emphasized, the decisive element within this system was the nexus between architecture and design: "It was the consolidation of design that revitalized the cultivated craftsmanship of the small series: objects inexpensive yet of high technical and design quality. This production was often 'anonymous,' making no reference to its design process but instead celebrating creativity, invention, and quality. At this juncture, the circularity between technique and art liberated many firms from stagnant provincialism, allowing some to establish branches worldwide, to collaborate with designers and artists, and to assume a social and cultural role as enterprises" (Vignozzi Paszkowski 2022, p. 124).

Complementing this view, the history of Made in Italy can be traced not only to large-scale industrial systems but to a dense ecosystem of small workshops, medium-sized firms, cultural salons, and galleries (Bosoni 2006, p. 31). In these informal contexts, experimentation could thrive: creativity was nourished beyond institutional boundaries through close ties among artists, artisans, and designers and through the permeability between artistic practice, entrepreneurial initiative, and cultural debate. Far from being peripheral, this fragmented milieu generated the inventive energy and flexibility that large industries could not sustain. Among its most vital nodes were galleries and workshops, conceived as unconventional spaces of display and encounter, where many of the most emblematic postwar initiatives first took shape.[5]

A striking example of this interplay emerged in Florence. In 1945, Danilo, Sergio, and Valeria Santi founded the Galleria Vigna Nuova, conceived as a space

[5] On recent studies concerning unconventional places and the strategies employed to preserve and disseminate their memories, see the essay by Bollati and Collina in this volume.

Fig. 4 Keystone Press Agency, Clare Boothe Luce, US Ambassador to Italy, at the Florence Artisan Fair, where an exhibitor presents her with a straw bonnet, 1955. Milan, Private collection

where craftsmanship, design, art, architecture, and publishing—most notably the journal *Esperienza Artigiana*—intersected. With architect Giovanni Michelucci's support, the gallery embodied the idea of craftsmanship as a vehicle of social renewal, and for two decades (until the devastating flood of 1966) fostered experimentation (Cozzi 2022). By 1947, this spirit soon found a broader stage in the relaunch of the *Mostra dell'artigianato*, which projected the city's artisanal tradition onto an international platform, aligning cultural heritage with Italy's export ambitions (Fig. 4).

In Rome, this innovation climate found expression in the Studio of Villa Giulia, established in 1944 by mosaic master and entrepreneur Enrico Galassi. Financed by Cesare Brustio and Aldo Borletti—owners of La Rinascente department store—the workshop convened a heterogeneous circle of artists and artisans, from Pietro Consagra, Leoncillo Leonardi, and Mirko Basaldella to Carlo Carrà, Giorgio de Chirico, Marino Mazzacurati, and Alberto Savinio. It became a site of experimentation across a broad spectrum of practices, including mosaic, wood inlay, marble, ceramics, and embroidery, with production destined for distribution through APE. Although critically acclaimed when presented in 1946 at Pietro M. Bardi's Studio d'Arte Palma,[6] the venture soon collapsed under financial strain, closing in 1949 (de Guttry and Maino 1994).

The atmosphere of those years—simultaneously fragile and brimming with possibility—was memorably described by sculptor Pietro Cascella, who recalled it as "a period of hope. Hope lingered in the air, tangible as stone". It was in this same spirit that Gaspero Del Corso and Irene Brin advanced their cosmopolitan vision through the Galleria L'Obelisco. Emerging in 1946 from the cultural milieu of the

[6] Bardi sought to export these art productions to South America, presenting them as a synthesis of Italian taste and style.

bookshop La Margherita, which had already convened figures such as Renzo Vespignani, Alberto Burri, Luchino Visconti, and Renato Guttuso, the gallery articulated a new model that combined talent-scouting with cultural mediation. L'Obelisco promoted Italian artists—including Giorgio Morandi, Mario Sironi, Giorgio de Chirico, Giacomo Balla, Massimo Campigli, Afro Basaldella, Giuseppe Capogrossi, Lucio Fontana, Alberto Burri, Arnaldo Pomodoro, and Corrado Cagli—while also introducing to Italy leading exponents of the European and American avant-garde. Exchanges with the Museum of Modern Art in New York and the Brooklyn Museum between 1948 and 1953 consolidated its role not merely as an exhibition venue but as an agent of cultural diplomacy and a mediator of Italy's identity within a transatlantic art world (Schiaffini 2018).[7]

Taken together, these experiences testify to the vitality of a cultural system where actors of different backgrounds and career stages collaborated to reimagine production for a new world and, in a sense, a new society. They functioned as laboratories of exchange, where the boundaries between art, craft, design, and commerce were constantly negotiated. Out of this fertile environment emerged the foundations of Made in Italy as a network of practices in which creativity and craftsmanship formed the basis of cultural identity and international projection.

4 From Constraint to Emblem

The reconstruction of Italy's international reputation after World War II cannot be reduced to the language of trade balances or industrial output. What defined the Italian case was the capacity to weave culture into production, so that goods circulated not only as commodities but also as sign-bearers of tradition, imagination, and beauty. Exhibitions, consortia, and galleries functioned as ancillary platforms and laboratories where local materials and craft traditions were reinterpreted and made legible to different audiences. Out of precarious circumstances emerged a system in which the "Made in Italy" label condensed centuries of artistic practice into the promise of modern desirability.

As economist John Kenneth Galbraith observed, the "miracle" of Italian recovery lay precisely in incorporating culture into everyday objects, so beauty became an economic resource. This insight has been developed further by philosopher Nicola Emery, who suggested that the label operated as a secular relic that transforms labor, memory, and tradition into a symbolic token of identity. To purchase an Italian object was, and remains, to acquire not merely a product but a fragment of

[7] The gallery introduced to Italian audiences major international figures such as Roberto Matta, René Magritte, Arshile Gorky, Marc Chagall, Wassily Kandinsky, Henry Moore, Alexander Calder, Salvador Dalí, Francis Bacon, Robert Rauschenberg, Balthus, Pablo Picasso, Wifredo Lam, Saul Steinberg, and Jean Dubuffet. It should also be noted that it was involved in organizing exhibitions in the United States, including *Twenty Imaginary Views of The American Scene by Twenty Young Italian Painters* (1953) and *Eterna Primavera. Young Italian Painters* (1954–1955).

cultural belonging—a dream of continuity with the *Bel Paese*, or the Beautiful Country, a traditional epithet for the nation (Emery 2024).

In this sense, the postwar affirmation of Made in Italy was neither fortuitous nor purely strategic: it was the articulation of a model where diplomacy, industry, and artistry coalesced. What had begun as a bureaucratic imposition—the compulsory stamping of "Made in Italy" on every exported item—was gradually transformed into a mark of distinction. The inscription that once burdened producers came instead to embody a cultural surplus, a visible guarantee that objects carried traces of creativity, tradition, and identity. This shift proved decisive for Italy's rehabilitation. It laid the foundations on which, in the following decades, Italian design attained global recognition as both a commercial force and a cultural narrative.

Competing Interests The author has no conflicts of interest to declare that are relevant to the content of this chapter.

References

Altea G (2024) Gio Ponti, la collaborazione con le ditte Singer e Altamira e l'immagine postbellica del design italiano negli Stati Uniti. Ital Stud 78(4):467–484. https://doi.org/10.1080/00751634.2023.2211494

Bedarida R (2023) Ceramiche per ricostruire l'Italia: Lucio Fontana nelle mostre americane del dopoguerra. In: Cordera P, Faggella C (eds) L'Italia al lavoro: un lifestyle da esportazione. Bologna University Press, Bologna, pp 107–124

Beltramini GM (ed) (1949a) Appunti per l'esportatore Italiano. The Monthly Review of the American Chamber of Commerce for Italy, vol I(1), pp 23–25

Beltramini GM (ed) (1949b) Per esportare negli Stati Uniti. The Monthly Review of the American Chamber of Commerce for Italy, vol I(10), pp. 3–9

Beltramini GM (ed) (1950) Prodotti italiani richiesti negli U.S.A. The Monthly Review of the American Chamber of Commerce for Italy, vol II(4), pp 10–11

Bosoni G (ed) (2006) The Italian way. Design and avant-garde in the 20th century. Skira, Milan

Calzini R (1948) Artigiani nostri alla 49a strada. Corriere della Sera, 18 April

Casali V (2020) Una favola americana. Il carteggio transatlantico tra Josef Singer e Gio Ponti (1950-1979). Engramma 175:347–369

Casali V (2022) The modern movement in Italy: architecture and design, 1953-1958. Territorio 100:142–151. https://doi.org/10.3280/TR2022-100017

Cervelli I (ed) (1952) From the chamber. The Monthly Review of the American Chamber of Commerce for Italy, vol II(3), pp 31–36

Cordera P (2022) L'incantesimo della casa. L'arte e l'industria in vetrina. In: Costa S et al (eds) Storytelling. Esperienze e comunicazione del Cultural Heritage. Bologna University Press, Bologna, pp 221–234

Cordera P, Turrini D (eds) (2025) Made in Italy on display. America and the creation of the myth between art and design. Cimorelli Editore, Milan

Cozzi M (2022) Giovanni Michelucci e il pensiero organizzato tra gli anni quaranta e cinquanta. In: Branca M, Cozzi M (eds) Architettura, arti applicate e industrial design negli anni della Ricostruzione postbellica toscana (1944–1966). Edizioni ETS, Pisa, pp 61–75

de Guttry I, Maino MP (1994) Le arti applicate a Roma negli anni quaranta. I laboratori, gli artisti, gli imprenditori. In: Roma Sotto le stelle del '44: storia arte e cultura dalla Guerra alla liberazione. Zefiro Editrice, Follonica, pp 139–169

Emery N (2024) Made in Heaven, made in Heaven. That's what they say. In: D'Acunto G et al (eds) Prospettive del Made in Italy. Libria, Melfi, pp 39–47

Schiaffini I (2018) La Galleria L'Obelisco e il mercato americano dal dopoguerra alla fine degli anni Cinquanta. In: Caratozzolo VC et al (eds) Irene Brin, Gaspero del Corso e la Galleria L'Obelisco. Drago Publishing, Rome, pp 125–141

Vignozzi Paszkowski M (2022) Gli artisti e la ceramica: pezzi unici e piccole serie. In: Branca M, Cozzi M (eds) Architettura, arti applicate e industrial design negli anni della Ricostruzone postbellica toscana (1944–1966). Edizioni ETS, Pisa, pp 123–135

Made in Italy Written in the Stars (and Stripes): Italian Craft's Global Influence in Vintage Newsreels and Documentaries

Giampiero Bosoni and Marta Elisa Cecchi

Abstract This essay offers a captivating exploration of how visual storytelling mediums, such as television and documentaries have played a pivotal role in shaping perceptions of Italian craftsmanship, design, and culture on both national and global stages. Notably, the inclusion of *La Settimana Incom* (or The Incom Weekly), an Italian newsreel distributed in cinemas from 1946 to 1965, highlights the influence of these media in promoting Italian design to a broader audience. These platforms not only showcased the elegance and innovation of Made in Italy products but also intertwined them with Italy's cultural identity, creating a compelling narrative that positioned Italian design as a symbol of quality and sophistication. By examining these visual platforms, the essay reveals their lasting impact in solidifying Italy's design legacy and fostering cultural appreciation both within the country and abroad. Additionally, it explores how these media helped establish a global recognition of Made in Italy as a brand that continues to influence contemporary design discourse.

Keywords Italian design · Newsreel · Documentary · Craftsmanship · Lifestyle

1 Introduction

How did the established image of a certain "Italian way" in advanced craftsmanship, in a varied series of industrial models, particularly the more personalised ones, and in general in what would later be recognised as Italian *good design*, come about in the contemporary world, especially since the second half of the twentieth century? What visions and stories have contributed to building the idea of Made in Italy abroad and a specific, perhaps somewhat distorted, image within our own *Bel Paese*? This contribution aims to explore in particular the crucial role of visual

G. Bosoni · M. E. Cecchi (✉)
Politecnico di Milano, Milan, Italy
e-mail: martaelisa.cecchi@polimi.it

P. Cordera (ed.), *Unfolding Made in Italy (1948-1962)*, Research for Development, https://doi.org/10.1007/978-3-032-10118-1_2

media—documentaries and period films, newsreels, but also films in the broadest sense of the term from those years—which, although accompanied by other channels such as newspapers, magazines and specialist publications, played a decisive role, more than has been appropriately recognised to date, in shaping and spreading this perception. It is through these visual languages that a powerful imagery has taken root, capable of evoking and sublimating the value, "creativity", and uniqueness of Italian know-how, both within the country and on the international scene, with a special focus on the relationship with the United States of America.[1]

Through the widespread success of cinema, before the advent of television, these media did not merely extol the elegance and innovation of Made in Italy products, but also knew how to intertwine these qualities, whether "new" or rediscovered and renewed, with a profound sense of Italian cultural identity through the persuasive power of the cinematic message. This new form of communication succeeded in constructing a "collective narrative" in which Italian "design" became an expression of values such as quality, refinement and modernity, an image so intense and captivating that it became part of the very life of the average Italian, the more or less conscious (so to speak "cinematic") protagonist of this period of great innovation, but also of progress that was as strong as it was full of contradictions, as Italian neorealist cinema soon highlighted with biting rawness.

2 Italian DNA

It is worth pausing to clarify what is meant today, and what was meant in the past, by "Made in Italy", an expression that gradually, and particularly since the 1980s, has transformed from a simple indication of geographical origin into a genuine cultural concept, capable of representing Italian quality, elegance, and manufacturing culture throughout the world. A *parola valigia* or portmanteau word (Dellapiana 2022) or a *label* encompassing different products, confers prestige. Although many brands had been using the Made in Italy label since the early 1930s, as Belfanti observes, the shift from "Made in Italy" as a simple label of origin where goods were produced into a symbol of a nation associated with cultural, aesthetic, and productive excellence occurred relatively late. Until then, especially in Italy, the prevailing trend was an uncompromising search for an authentically Italian language in how things were done, in products and manufactured goods, an identity built by pioneering and courageous companies that wrote the first pages of the history of Made in Italy. According to Corbellini and Saviolo (2004)—also cited by Belfanti (2019)—this Italian productive culture draws its lifeblood from what can be considered the true *DNA* of Made in Italy: an authentic "*Renaissance effect*", where it is not only the search for beauty that counts, but also "del ben fatto", or

[1] On the promotion of Made in Italy through exhibitions, see Cordera's essay in this volume.

well-made.[2] In this sense, linking the Renaissance as a key to understanding the culture of Made in Italy, it is particularly interesting to refer to the "Italian identity" discussed by Enzo Biffi Gentile (1995), but above all to the *Modo Italiano*, or the Italian way (Bosoni 2006) that indicates that specific Italian attitude towards design, deeply rooted in historical culture but at the same time projected towards innovation, which represents in all respects the true cultural heritage of Italian design.

Returning to the analysis of the visual platforms under consideration, it is clear that they have exerted a decisive influence in consolidating a coherent image of Italian "style", "taste" and "way", within which the nascent culture of design has developed and articulated itself. These media have contributed to structuring a recognisable aesthetic imagery and promoting widespread cultural appreciation, both nationally and internationally. In this context, the emergence of the Made in Italy concept has taken on a meaning that transcends the mere geographical indication of origin, becoming an identity brand capable of synthesising design quality, innovation, and shared symbolic values. Made in Italy has thus established itself as a strategic element for the commercial expansion of Italian products in foreign markets and for strengthening a sense of belonging and collective pride at home. This dynamic became particularly significant in the postwar period, especially in the early years of reconstruction, a period marked by a drive towards the future, a renewed desire for participation, and a substantial investment in design as a form of rebirth and social cohesion. Furthermore, the Italian designer Vittorio Gregotti, in his renowned publication titled *Il disegno del prodotto industriale. Italia 1860–1980* (1982) stated, "It should not be forgotten that, starting in the 1950s, Italy's industrial economic development underwent a sudden acceleration. Despite the continuing regional income imbalances, this opened the gates to a series of long-awaited aspirations in terms of services and consumer goods. With the end of the conflict and national liberation, the destroyed cities, the "sigorine" (young women), the American invasion of Italy, chewing gum, powdered milk, Coca-Cola and, above all, the ideology of "comfort" home mechanisation had arrived: the myth of the refrigerator, of household appliances and, above all, the competitive desire to achieve well-being and security, seen as possible and therefore achievable examples had been created [...]".

[2]According to Belfanti, "beautiful and well made" encompasses a dual tension, aesthetic and design, which translates into the ability to work with and enhance the available materials and resources. This Italian attitude has its roots in the tradition of Renaissance workshops and guilds, fueling a distinctive know-how that, through the transformations of the postwar period, particularly in the 1950s, has survived to this day.

3 La Settimana Incom: The Newsreel

In a climate of recovery and enthusiasm after the end of the war and the fall of the fascist regime, but after the political settlement of 1948, the "official" message of the government projected a sugar-coated image that sought to portray widespread optimism and enthusiasm for a future perceived as bright and accessible to all social classes. *La Settimana Incom*[3]—an Italian newsreel distributed in cinemas between 1946 and 1965—is a case in point of the role played by such tools in promoting (a few years earlier, it would have been called propaganda) of this new way of making products, "goods", to be offered to a broad and diverse audience through a cheerful and persuasive image to be communicated to the world, especially the US. The message is clear: an authentic people, characterised by hard work and ingenuity, capable of facing difficulties with a smile, who value craftsmanship and simple solutions, creating products with local resources, but always original, sometimes eccentric, and distinguished by an authentic and genuinely popular spirit. Beyond rhetoric and obvious problems, this picture is entirely acceptable, and it is precisely in this context that Italian design will gradually become more aware of itself and the changed international context. The aim was not only to promote the commercial expansion of national products, but also to support the image of redemption and progress that postwar Italian society felt was necessary to rebuild its identity after the deep divisions caused by the World War II.[4] In this context, institutions such as CADMA (Commissione Assistenza Distribuzione Materiali Artigianato, *or* Commission for Assistance and Distribution of Handicraft Materials) and APE (Artigianato Produzione Esportazione, or Handicraft Production Export) played a decisive role in promoting and distributing popular Italian handicrafts overseas. It is significant to note that, in selecting the protagonists of this narrative, preference was often given to figures who were less central to the national scene but perceived as more responsive to the foreign imagination of Italian taste and "style". Just think, for example, of the comparison between Franco Albini, one of the most important Italian architects and designers, capable of dialoguing with American trends, and Fabrizio Clerici, a painter and set designer well known in the art world but less valued from a historiographical perspective, whose figure was nevertheless more easily marketable on an international level. To further complicate this process, Ghianda (2022)

[3] The programme was launched after the Second World War on the initiative of Sandro Pallavicini, founder of the INCOM production company, together with journalist Luigi Freddi, to counter the monopoly of the Istituto Luce. Pallavicini was its director until 1956, overseeing the entire newsreel activity. Initially financed by Italian Swiss capital (Cedraschi), the project gradually took on a pro-government orientation, especially after the growing involvement of the Christian Democrat party from 1948 and following the legislative measures of 1947 and 1948 in support of cinematographic activities.

[4] Emblematic of this type is Carosello, a television advertising programme broadcast by Rai, conceived as a container for short films commissioned by companies to promote their products. Launched on 3 February 1957, Carosello was broadcast daily in the evening, immediately after the news, until 1 January 1977.

notes, there was the historical backwardness of the Italian industrial system, which was nevertheless partly offset by small-scale craft and mechanical prototyping production (cars, airplanes, motorcycles), which is also evolving into updated forms of mechanisation, supported by an ancient and never-ending preference for high-quality 'handmade' products. While this attitude initially limited industrial development, it also enhanced the extraordinary quality of Italian products, making them unique on the world stage.

The episodes of *La Settimana Incom*, lasting an average of about 10 minutes, were screened before the start of the films and mainly dealt with issues related to the reconstruction of a country deeply scarred by the destruction of war. Like other newsreels produced by INCOM (Industria Corti Metraggi Milano) (Fig. 1), *La Settimana Incom* sought to spread a shared national narrative, but it also tended to conceal the deep social inequalities that consolidated in postwar Italy, thus providing a partial and somewhat sugar-coated representation of reality.

However, this move should also be interpreted as a deliberate communication strategy, capable of modulating language to make it appealing to domestic and international audiences. This was not necessarily a superficial sensitivity, but rather a functional approach to capturing folkloric and stereotypical aspects considered interesting by foreign observers, to facilitate the entry of Italian products into global markets. The light, positive, and undramatic language was in line with the rhetoric typical of the so-called "American dream", evoking a welcoming, optimistic, and exciting image, capable of conveying the idea of a future that could be built from the bottom up. In this sense, adopting communication models linked to entertainment (characterised by lively, captivating, and sometimes spectacular tones) responded to specific market logic. It helped to promote an image of Italy consistent with foreign expectations. From this perspective, the choice to emphasise certain types of industrial and craft products and the emerging "Italian good design" (sometimes marginal in the national context but successful elsewhere) was a targeted and conscious strategy aimed at promoting.[5]

[5] On the other hand, the Istituto Luce (L'Unione Cinematografica Educativa), founded in 1924, was created as a tool for political propaganda, aimed from the outset at shaping public opinion through a selective representation of reality. During the Fascist regime, it was one of the main vehicles of official visual communication, particularly through the newsreel *Giornale Luce*, distributed regularly from 1927 onwards. During the 20 years of Fascism (1922–1943), *Giornale Luce* became an effective tool for spreading Mussolini's ideology, shown in cinemas before screenings and aimed at celebrating the regime's achievements. Short films of 8–10 minutes, constructed with accessible visual language, solemn music and rhetorical narration, helped to reinforce the image of the efficiency of the fascist state and consolidate the cult of the Duce, systematically obscuring dissent. With the fall of Fascism, the Institute was forced to redefine its function. In the postwar period, while maintaining an institutional slant, newsreels adapted to the new democratic context, expanding their content to include cultural, scientific and current affairs topics, with a more sober style that was closer to reality. The rise of television in the 1950s and 1960s marked the gradual decline of newsreels as a central medium of audiovisual information.

Fig. 1 Header of the La Settimana Incom newsreel, *Due civiltà s'incontrano*, 1957. Retrieved from Istituto Luce Archive, Rome

3.1 Made in Italy

An emblematic and early example of this Italian culture narrated through film is the short black-and-white documentary in the Istituto Luce online archive, *Made in Italy. Fabbricato in Italia, or Manufactured in Italy* (1953), directed by Renato Tonini,[6] with photography by Emanuele Lomiry (a photographer with a futurist background and professionally linked to the Milan Trade Fair), Arturo Climati, Victor De Sanctis, Lorenzo Fiore, and Antonio Marzari, all professionals already active in Cinecittà in the early 1940s. The 7-minute short film stands out for its lively and dynamic editing, accompanied by the characteristic music of newsreels of the time.

The work is based on the remarkable fact that in 1952, almost five and a half million foreign tourists visited Italy on holiday to enjoy its landscapes and monuments. However, the narration is entrusted to Guido Notari's unmistakable voice,[7]

[6] In the late 1950s and early 1960s, he was an important executive producer of renowned Italian films by Corbucci, Steno, De Martino and Camerini, with actors such as Totò, Peppino De Filippo, Alberto Sordi, Walter Chiari and Vittorio De Sica.

[7] In the 1920s, he worked in commerce until becoming advertising director at la Rinascente department store. Gifted with a clear voice, free of dialectal inflexions, he participated in the announcer competition organised by the Italian Radio Auditions Agency (EIAR). He won the competition and

Fig. 2 Giulio Minoletti, Front of the lead carriage of the ETR 300 train, known as Settebello [One's best card], produced by Breda Ferroviaria, 1952

famous speaker of *La Settimana Incom* since its foundation in 1946, immediately takes an alternative route, aimed at highlighting what he defines as "the primacy of primates that the world recognises us for, that of being the first in Europe to reach the goal of national reconstruction" (Istituto Luce 1953). The documentary thus offers a rapid succession of emblematic images: from the famous panoramic carriage of the "super-electric train speeding at 160 km/h", which is the brand new ETR300 of 1952, known as Settebello (Fig. 2), designed by architect Minoletti for the Breda company, to Roe's Termini station, the "largest station in Europe", inaugurated in 1950 and designed by architects L. Calini, M. Montuori and V. Vitelozzi ("as for the façade, many things have been said, to each his own, we like it!"), to the Stadio dei Centomila,[8] with the 1960 Olympics renamed Stadio Olimpico—defined as a "combative emblem of vital competition"—to the steelworks on the seashore at Cornigliano, defined as "architecturally beautiful" (sigh!) among the most modern in Europe.

The AGIP methane pipeline facilities sought and desired by E. Mattei in Cortemaggiore are shown, whose derivatives feed the manufacturing industry for major trade fairs, such as those in Milan ("the largest trade fair in the world", and you can see the Palazzo delle Nazioni designed by architects A. Bianchetti and

was hired as chief announcer, first at the Milan office and then permanently in the capital, where he remained until 1943, continuing to work for RAI.

[8] Restored by architects C. Valle, C. Roccatelli and A. Vitelozzi, based on an earlier design by E. Del Debbio and L. Moretti.

C. Pea, built in 1948) and Turin, Verona and Bari. A FIAT tractor, with a slim, compact design and a small size suitable for Italian terrain, is presented in this context. A section is dedicated to the military sector, which is significant considering that Italy, having lost the war, had severe limitations on the production of war material, so much so that an OM truck for troop transport is presented, which is very versatile on rough terrain, "built in just 4 months and adopted by NATO armies". This is followed, still in a lively military vein, by the presentation of the famous Italian car manufacturers Alfa Romeo and Ferrari, which dominate the international Grand Prix scene. Particular attention is paid to city cars designed by famous Italian coachbuilders, praised for their "sleek lines without frills or paradoxes, striking such a perfect balance between functionality and harmony that Americans come to our coachbuilders to have their engines and chassis dressed", in this case, first showing a model of the Nash-Healey,[9] redesigned by Pininfarina, followed by the rear of the spectacular 1952 Chrysler d'Elegance with its automatic spare wheel, designed by the director of the Chrysler Style Centre, Virgil Exner, together with Italian coachbuilder Ghia, who produced it in Italy, and with the consultancy of Giovanni Savonuzzi. The filmic rhythm slows down to accommodate the almost theatrical slowness of the models wearing Italian dresses—"Dress Italian! The trend spreads parallel between car manufacturers and ladies"—introducing national fashion and its renowned elegance. In a dizzying thematic leap, the narrative shifts to large dams, such as the Santa Giustina dam in Val di Non, one of the highest in Europe, inaugurated in 1951 and designed by engineer Claudio Marcello, while the hydroelectric power station was designed by Gio Ponti, strategic at the time for electricity production, and racing boats, with the hull Laura I with an Alfa Romeo engine (the same that had led drivers Farina to victory in the Formula 1 World Championship in 1950, and Fangio in 1951) built by Guido Abate and driven by the famous driver Mario Verga (a well-known silk entrepreneur) for the world speed record. As it races across Lake Garda at Campione, the narrator exults, evoking "the waters that seem to rebel against the slap of speed". The film continues with the great names of Italian motorcycling around the world, from small to large displacement (the 1952 MV Augusta 125 cc monoalbero corsa probably ridden by Cecil Sandford, the Moto Guzzi 250 cc Gambalunghino ridden by E. Lorenzetti, and the Gilera 500 cc. 4C, designed by Piero Remor and ridden by U. Masetti), capable of winning all the world championships, and with the bicycles of Gino Bartali, who raced on Legnano bikes, and the solitary Fausto Coppi, riding Bianchi bikes, both with Campagnolo gears. From Coppi's "tireless heart", we move on to the "first artificial heart in regular use" created by Dr. Mario Dogliotti, who pioneered cardiac surgery in Italy. Then, in sequence, the textile machines skilfully tuned by Italian weavers to achieve a truly

[9] In 1952, Nash commissioned Pininfarina to restyle the model. The updated version was presented in the same year at the Chicago Auto Show. Following this agreement, the Nash-Healey began to be completed at Pininfarina's factory in Turin.

unique "feel", so much so that, as the newsreel tells us, Italian fabrics were so sought after that even prestigious foreign brands sometimes appropriated them, affixing the most renowned label of the time, "Made in England". Then came hand-painted ceramics from Salerno, exported to American department stores, and their hallmark was Made in Italy. These were followed by Olivetti typewriters designed by Marcello Nizzoli (Lexicon 80 and Lettera 22 typewriters, Elettrosumma MC14, Multisumma 14, and Divisumma), displayed in the Olivetti shop in Via del Tritone in Rome, designed by Ugo Sissa and with Renato Guttuso's famous *Boogie-Woogie* painting in the background. "The double-cylinder machines that produce men's nylon stockings abroad, churning them out like a long lasagna, are made in Italy."[10] We then move on to the film industry, with a fast pace accompanying images of Cinecittà and its divas, such as Sophia Loren and Anna Magnani "who is an Oscar winner, as is Vittorio De Sica, who challenges the film with his incendiary cigarette!". This is an opportunity to introduce the film and theatre production theme with energy and wonder. We then move on to tobacco production, the bane and delight of smugglers and smokers, whose leaves, "destined to vanish in smoke", symbolically prelude the floral theme of the exporting over 265 million roses and carnations yearly. From flowers, we move on to fruit, particularly oranges, poetically described as "rays of sunshine frozen in sweet pulp, wrapped in aromatic golden peel", which will become preserves and jams in tins in Italian homes and beyond.

The lyricism of these descriptions finally opens up to music, typically Italian, evoked by the notes of the most popular product, the accordion, "198,000 of which crossed the border in 1952", before moving on to the lively street noise of the Vespa, "the most practical way to put a man on a motorbike" (incidentally, a low-floor body designed especially for more agile female use), a product of Piaggio's industrial conversion from war production, the symbol par excellence of Made in Italy, patented in 1946 and designed by aeronautical engineer Corradino D'Ascanio. At the end, as we see many boxes bearing the Made in Italy mark being loaded onto ships, we also see an elegant Alfa Romeo 1900 Sprint spider being loaded, with mechanics designed entirely by engineer Orazio Satta Puliga and bodywork by Touring Superleggera.

The film ends with a nostalgic reference to the theme of emigration. However, it immediately realigns with a reflection on exports—and packaging ("trains run, planes fly, ships hum under pressure. Why? Eh, because there is a label that's making its way... Made in Italy!"), recalling that more than 189 billion per year fuel an impressive traffic flow on all the roads, "that freedom has reopened to us".

[10] Bruno Munari was inspired by the discovery of this type of seamless cylindrical yarn production in Italy to design his Falkland lamp, produced by Bruno Danese in 1964.

3.2 Due civiltà s'incontrano

Another film, this time in colour, found in the online archives of the Istituto Luce, focuses solely on the relationship between Italy and the United States. Entitled *Due Civiltà s'incontrano,* or Two Civilisations Meet, circa mid-1950s (probably 1957), it was directed by Vittorio Gallo[11] and, with cinematography by Emanuel Lomiry (the same as the previous film), and lasts 14 minutes. The sequence of content and its dynamic and engaging editing echoes that of *Made In Italy;* however, in this case, the overall tone appears more poetic and allusive, trying to emphasise parallels and influences between the two countries. These parallels are real but imbued with an almost prophetic aura, as if they were written in the stars and destiny. In this sense, the film's beginning is particularly significant, parallel to the US history, whose origins are evoked by "an Italian name, Christopher Columbus" (Istituto Luce 1957).

The film continues by emphasising how numerous Italian surnames are in the United States and how, just by taking a few steps, one can perceive, see, or imagine Italian presences in many cities, with a red Alfa Romeo Giulietta Spider[12] and a Fiat 600[13] speeding by in the background. "In simple terms, Italian taste is in vogue". The foreign press is also in constant contact, exchanging information "as if we were in two cities separated by only a few kilometres".

Interest in Italy also extends to the language, books and newspapers (various newspapers are displayed in a newsstand, including *Il Corriere della Sera*, *Il Mondo*, the weekly magazine *Epoca*, the magazine *Il Settimo Giorno*, *Il Tempo*, and also *La Settimana Incom Illustrata*)[14] and many other cultural sources, not to mention shops offering innovative design products and original Italian artefacts. Among the images, for example, is the famous Olivetti store in New York, designed by the BBPR studio with bas-reliefs by sculptor Costantino Nivola, where the typewriters designed by Nizzoli stand out, placed on elegant stalagmite-like supports at the entrance, and the movement of the wheel inside the display (Fig. 3).

The story then moves on to Italian fashion,[15] followed by images dedicated to Murano glass, before moving on to another icon: the Vespa, a symbol of freedom and lightness. These suggestions lead to the conclusion that, ultimately, "Americans

---

[11] A documentary filmmaker who has made numerous works since the 1940s, including the Astra documentary series (1954–1960), the documentaries *Emilia* (recounting the effects of the Marshall Plan economic aid in the region), *Casello 77, Acciaio* e *Disco Verde per il Sud* (1956), which featured the exclusive ETR 200 train known as the "viper's head", produced in the Breda railway workshops in 1936 and designed by Giuseppe Pagano with interiors by Gio Ponti.

[12] Built by Italian car manufacturer Alfa Romeo from 1955 to 1962.

[13] Designed by Dante Giacosa, built between 1955 and 1969, it is considered the icon of the Italian economic boom.

[14] Represented by Michelangelo's David, a publication linked to the newsreel of the same name that is the subject of the analysis.

[15] Also interesting is the 1955 film *Allo specchio del mare. Su transatlantici americani lungo la "rotta del sole" lusinghiere immagini della moda italiana*—or In the mirror of the sea. On American ocean liners along the "sun route", flattering images of Italian fashion—are also in the

Fig. 3 Olivetti shop in New York, 584 Fifth Avenue, designed by BBPR in 1954, in the Incom newsreel, *Due civiltà s'incontrano*, 1957. Retrieved from Istituto Luce Archive, Rome

become Italianised". The whole scene is set in Orsini's café,[16] where, among the servers, it is possible to catch the classic La Pavoni (Fig. 4) espresso machine, still in the version with a vertical boiler, when the same company began producing La Cornuta D.P. 47, designed by Gio Ponti, with a horizontal boiler, in 1947.

The article then moves on to contemporary art and American galleries, which increasingly welcome Italian artists and works, including paintings by Massimo Campigli. The article then mentions universities, from Italian literature courses at Columbia University to the presence of Italian students, "children of our emigrants", at Harvard, where Italian culture seems to be returning to "the place of honour it has always deserved" as a sign of redemption and a desire to be present in the contemporary world. No less significant is Italy's contribution to science: for example, Professor Rossi, a respected collaborator of Enrico Fermi, appears, directing research on cosmic rays before moving on to medicine and research into cancer treatment.

The climate of war rhetoric accompanies the sequence dedicated to aviation, with young Italian pilots in Texas representing the new generation and future promise of the Air Force.

---

online archives of the Istituto Luce, which show fashion shows and clothes on American ocean liners bound for Italy.

[16] Located on 56th Street in New York and run by Armando Orsini and his brother Elio, Orsini's was a popular venue for actors, actresses and various personalities, famous for its glamorous dolce vita atmosphere.

Fig. 4 Gio Ponti, espresso coffee machine, known as La Cornuta [Horned], produced by La Pavoni, 1949

The documentary then moves on to politics, with views of Washington, mentioning the extensive collections of modern Renaissance paintings and how many descendants of Italians now hold prominent positions at the United Nations, "making their voices heard authoritatively". A reflection on trade follows this, with images of the ocean crossed every day by ships laden with Italian goods arriving in New York. The tourism theme is then introduced, recounting how many travellers choose transatlantic liners to reach their favourite Italian destinations, immediately breathing in "a typically Italian atmosphere, made up of courtesy and affable cordiality". First, the turbo liner Cristoforo Colombo appears in the distance,[17] while the

[17] It was an ocean liner belonging to the Genoese shipping company Italian Line (1953–1983), sister ship to the Andrea Doria, but with an integrated alternating current electrical system and a different interior layout. In the spring of 1964, the Cristoforo Colombo transported Michelangelo's Pietà from the Vatican to the World's Fair in New York that year.

images of the sun deck with the swimming pool and bathers around it refer to the ship Conte Grande.[18]

The film ends in Italy, at the University for Foreigners in Perugia, where many people learn Italian. It then returns to music, with the image of a young pianist—reminiscent of young singer Nina Simone—and prestigious scholarships for the arts, such as those offered by the American Academy on the Janiculum Hill in Rome, rekindling the theme of Italian creativity. Ultimately, the story returns to the United States, with its busy libraries housing an extraordinary wealth of books, before mentioning the development of construction techniques in Italy, featuring the Breda tower skyscraper[19] and the building in Piazza Missori in Milan, and industrial production, showing the assembly stages of the Lambretta (125 LD model in the 1953 fairer version).[20] Moreover, the Fiat 1100/103 supermini cars of 1953 and the 600 of 1955, designed by Dante Giacosa, and the assembly of the first G91 Fiat-Aviazione light tactical jet fighters of 1956, designed by engineer Giuseppe Gabrielli and adopted by NATO in 1958, parallel to scientific progress, to which numerous Italian scholars contributed. The film ends with a shot of an Alitalia aircraft taking off, symbolically projecting itself towards new horizons.

4 Conclusion

In the complex landscape of postwar reconstruction in Italy, newsreels played a crucial role not only in documenting the changes taking place, but also in constructing an "ideal" and, in some ways, strategic narrative of what was intended to be Italy's new "destiny", which seemed to project it towards a future of excellence in manufacturing, in the new culture of design and in the cultural vision that the term Made in Italy could generally embody. These tools, capable of capturing the collective imagination through the seductive language of cinema, actively contributed to shaping it, in the most profound and most strategic sense of the term, but also to spreading that image of Made in Italy which, over time, would establish itself as a globally recognised "brand identity".

Long before the advent of television, cinema—and in particular the short films shown before feature films—managed to convey with extraordinary effectiveness an idea of "Italianness" based on the rich and complex intertwining of inventiveness

[18] The Conte Grande, launched in 1927, made its maiden voyage to New York in 1928. From 1937, it was used for voyages to South America, where, at the outbreak of the war, it was seized and sold to the United States (later renamed Hermitage). Returning to Italy in 1947, it was in poor condition and without its original interior by Adolfo Coppedè, it was decided to refurbish it in 1949. A competition was held to design the first-class public rooms, which Gio Ponti and Giovanni Zoncada won (Averna 2025).

[19] Designed by Eugenio and Ermenegildo Soncini and Luigi Mattioni, before the construction of the Pirelli Tower, it was a symbol of Milan's renewal and recovery after the Second World War.

[20] Produced by the Innocenti mechanical engineering company in Milan from 1947 to 1971.

and craftsmanship in a typically Italian way, based on the eternal historical condition, clearly evident during the early postwar period, of "fare di necessità virtù"—or make a virtue out of necessity—as Gio Ponti often repeated.

It is precisely in this convergence between communication, identity and the precursors of "Italian-style" design that the roots of the Made in Italy model lie: a cultural projection that took shape thanks to a narrative intuition that was consolidated through media tools such as newsreels, spreading throughout the world as a curious and unusual synthesis of tradition and innovation capable of translating into success. For this very reason, it is essential for our historical research to analyse this type of popular media material (think also of the magazines, photo novels and comics of this period), recognising it as a different and additional resource compared to traditional ones, and thus discovering its particular interest for study, enabling us to understand even better the value and social context of design culture. It is necessary to develop new methods and new tools for reading, studying, and analysing these particular documents correctly, both in terms of their explicit content and the rhetorical and emphatic language (in some cases almost poetic) used, to fully understand what can be read between the lines: the system of values, aspirations and visions of the future that they conveyed.

Only through careful reading, capable of grasping both the communicative and aesthetic-rhetorical dimensions, can these materials reveal their nature as unique historical documents, shedding original light on the cultural and design roots of Made in Italy.

Competing Interests The authors have no conflicts of interest to declare that are relevant to the content of this chapter.

References

Averna M (2025) Conte Grande. In: Transatlantic transfers: atlas. Politecnico di Milano, Milan. https://transatlantictransfers.polimi.it/it/atlas/69/conte-grande/. Accessed 19 Jul 2025

Belfanti CM (2019) Cultural history of made in Italy. Il Mulino, Bologna

Biffi Gentile E (ed) (1995) La Sindrome di Leonardo. Arte Design in Italia 1940–1975. Umberto Allemandi, Turin

Bosoni G (ed) (2006) The Italian way. Design and avant-garde in the 20th century. Skira, Milan

Corbellini E, Saviolo S (2004) The made in Italy challenge. Etas, Milan

Dellapiana E (2022) Design and the invention of made in Italy. Einaudi, Turin

Ghianda ML (2022) The true history of made in Italy. Doppiozero. https://www.doppiozero.com/la-vera-storia-del-made-in-italy. Accessed 19 Jul 2025

Gregotti V (1982) In: Bosoni G et al (eds) Il disegno del prodotto industriale. Italia 1860–1980. Electa, Milan

Istituto Luce (1953) Made in Italy (fabbricato in Italia). Archivio Storico Istituto Luce. https://patrimonio.archivioluce.com/luce-web/detail/IL5000025850/2/-11029.html. Accessed 19 Jul 2025

Istituto Luce (1957). Due civiltà s' incontrano. Archivio Storico Istituto Luce. https://patrimonio.archivioluce.com/luce-web/detail/IL3000089198/1/-2399.html. Accessed 19 Jul 2025

Icons of Memory and Modernity: Italian Hotels of the 1950s and the Promotion of Italian Lifestyle

Francesco Scullica and Elena Elgani

Abstract This essay examines the role of Italian hotels built in the 1950s as key promoters of Made in Italy design and lifestyle. These hotels, designed and furnished by iconic Italian designers, served as showcases of Italian craftsmanship and design nascent industry during the period of Reconstruction. Focusing on how interiors and spatial narratives, as well as the system of product-services, contributed to constructing a modern image of *leisure*, in particular during tourism experiences, supporting the affirmation of Made in Italy, the paper underlines the centrality of architect Gio Ponti. Through projects and writings, particularly the Parco dei Principi Hotel in Sorrento, blending research and innovation, he redefined hospitality spaces as a dynamic field of experimentation and as expressions of Italian creativity and *savoir-faire*. Reflecting a modern, aspirational Italian lifestyle to international guests, the hotels contributed to spreading values of *Italianess* and a desirable Italian lifestyle. Today, these hotels remain icons, bridging past and present by preserving their legacy while continuing to embody the enduring timelessness of Italian design.

Keywords Design for hospitality · Tourism · Interior design · Gio Ponti · Hotel Parco dei Principi

1 Introduction: Hotel Design Before the Second World War

The Second World War destroyed much of the tourist infrastructure that had been built in the early twentieth century to support the development of tourism throughout Europe. In Italy, the Fascist period prioritised fast transport, such as cars, trains, and airplanes, and encouraged the regeneration of the "new man" through outdoor activities that connected people with nature and stimulated the body. Holiday

F. Scullica · E. Elgani (✉)
Politecnico di Milano, Milan, Italy
e-mail: elena.elgani@polimi.it

P. Cordera (ed.), *Unfolding Made in Italy (1948-1962)*, Research for Development, https://doi.org/10.1007/978-3-032-10118-1_3

destinations and regions that could compete with those in neighbouring countries, such as the Ligurian Riviera and the Dolomites Mountains, received more attention and were promoted by fascist propaganda. Many designers, influenced by the rationalist approach, saw hotels as an architectural typology to be reinvented in relation to construction innovations, the introduction of new technological and plant engineering solutions to improve the performance of spaces, and a rethinking of interior design in terms of equipment and comfort. New openings in Italy and in the colonies of the Regime, along with renovations of existing buildings, and conversions of historic buildings were widely documented in the pages of the Italian Touring Club's magazine *L'Albergo in Italia*, published between 1925 and 1943 (Kawamura 2021). This focus on accommodation facilities is exemplified by the exhibition on hotel furnishings organised for the 7th Milan Triennale in 1939 (Pasquali 1940). This exhibition was an important step in establishing a design approach for interior design in the hospitality sector. Many designers, including Guglielmo Ulrich and Giulio Minoletti, were commissioned to design different types of hotel rooms. During the 1930s, the study of hotel typology centered on classifying the features of accommodation facilities based on their location (e.g., by the sea, in the mountains, or in the city) and the duration of the stay (e.g., long-term, medium-term, or transit). The exhibition influenced the evolution of hotel furnishings in Italy by stimulating a design culture that was aware of the needs of "collective living" and by promoting the dissemination of new concepts through specialised magazines, such as *Domus* and *Casabella*. The rooms and furniture systems, designed according to criteria of modularity, standardisation and integration between space and product, demonstrate the new interest in hotels, which arrived in Europe from the United States after the end of the American Civil War and with the development of railway networks connecting the major American cities, associating the new approach in relation with the culture of the Modern Movement and in particular Rationalism.

During these years, architect Gio Ponti began to explore the design of hospitality spaces in two design projects for hotels on the Tyrrhenian and Adriatic coasts. These projects reveal a rationalist approach to hotel design, drawing on Le Corbusier's concept of the *machine à habiter* (Le Corbusier 1923) and applying it to hotel construction. Compared to the large historic Italian hotels and the widespread precarious accommodation facilities, Ponti conceived the hotel as a "machine for living", considering not only the functional and managerial aspects, but also the relationship with service areas—those that would be defined in US hotel design manuals as the "back of the house"—and internal distribution routes (Penner and Rutes 1985).

The destruction caused by World War II affected all hospitality facilities, which were often converted into hospitals or shelters for troops. It also affected all road and rail infrastructure, which enabled travel within and outside national borders. This almost completely interrupted tourist flows in Europe. By the end of the war, the only infrastructure left intact were the airstrips. Having long been used for military purposes, airstrips and airports were ready for rapid conversion to civilian flights, which could transport small groups of tourists, particularly from the United States (Battilani and Fauri 2020). However, air travel remained accessible only to a few wealthy travellers after the war, while most people did not have access to organised tourist experiences.

2 Renewed Interest in Hospitable Spaces in Postwar Reconstruction

The Reconstruction of Italy presented an opportunity for a comprehensive renewal of the entire tourism system, starting with hotels, thanks in part to aid from the Marshall Plan.[1] This renewal was not only about reconstruction, but also about researching new developments and new ways of understanding hospitality. In 1950, a Marshall Plan-funded Italian mission organised by the Economic Cooperation Administration (ECA) allowed architects such as Ignazio Gardella and leading hotel managers to study the North American tourism industry first-hand (Sintini 2018).

During this period, numerous transatlantic exchanges stimulated a renewal of the design approach to hospitality spaces, considering as in the same years, large hotel chains in the United States began building major hotel complexes, such as those in Miami, designed by Morris Lapidus. During the reconstruction period, tourist numbers gradually resumed their flow towards seaside and mountain destinations from Switzerland, France, and Germany, and were joined by a small number of tourists from the United States who were characterised by a greater interest in cultural sites and cities of art. Domestic tourist numbers also subsequently increased, stimulated by the reconstruction of the railway network, the availability of newly built motorways, and the spread of mass-produced cars. The opportunity to travel has increased the number of potential tourists and stimulated the design and construction of more affordable travel and accommodation infrastructure, as documented in the pages of *Turismo e Alberghi*. The magazine published by the Italian Touring Club since 1946 (De Falco 2019) has supported the promotion of certain Italian regions by highlighting their unique characteristics and raising awareness of their distinctive contexts (Berrino 2021). This has enabled tourist experiences to spread even among the middle classes, thus increasing the number of tourists.

The reconstruction process has indeed stimulated the proliferation of numerous small, family-run hotels, as well as more advanced, structured hotels, especially along the coasts and in mountainous regions, and in urban centres considered to be of cultural importance. The reconstruction and new construction of hospitality venues was accompanied by a reworking of consumer practices. The components of the tourism industry established in the 1930s were revamped with new images and scenarios, strongly influenced by American culture, which allowed tourism to be incorporated into the new narrative of the democratic state and economic development focused on the opening of markets (Battilani 2001).

[1] On the Marshall Plan in relation to exhibition and display, see Cordera's essay in this volume.

This aspect has a profound influence on the definition of the Italian approach to hospitality, which is closely linked to the concept of the Italian lifestyle. This way of life still influences the design of accommodation facilities today. However, it is not possible to see a smooth continuity in tourism design between the period before and after the Second World War. In fact, “while from a functional point of view, fundamental innovations in terms of infrastructure, means of transport and architectural styles had already been introduced, their symbolic meaning (soft innovation) was completely rewritten during the democratic period, and with it the entire tourist imagery” (Battilani 2020, p. 127).

In the 1950s, the gradual increase and spread of tourist experiences led to the emergence of mass tourism in Italy. A significant number of workers were now able to enjoy paid holiday time and began to take holidays in seaside or mountain destinations. The availability of free time from work allowed for the exploration of new social times, rituals, and spaces. Hospitality spaces, as well as the entire tourism infrastructure system, acquired new semiotic content related to freedom of movement and the pursuit of happiness and well-being. New behaviours and new consumption models emerged, defining a new generation of Italians who were more attentive to leisure. Free time is increasingly becoming a time for consumption (Corbin 1995), self-expression, and participation in a collective dimension, other than the domestic and every day one. Consequently, there has been a resurgence of interest among designers in the field of accommodation facilities. Indeed, since the postwar period, and particularly from the 1950s onwards, there has been a codification of an approach to hotel design that is more attentive to experiential aspects, which are now defined as “lifestyle”. The experiential aspects of new ways of living concern the experience of rest, relaxation, refreshment, and relationships within hospitality facilities. The concept of functionality and structural efficiency is thus intertwined with notions such as aesthetic, emotional processes and experiential appeal (Deambrosis and De Magistris 2015) as underscored by Ponti himself in the article dedicated to the swimming pool he designed for the Hotel Royal in San Remo (Imperia): “not a new swimming pool on the Riviera, but a new attraction among the pools of the Riviera; this is the functionality, and the true and serious issue” (Ponti, p. 12).

In addition to the renovation and adaptation of existing historic hotels, new types of accommodation were also developed, particularly mid-range establishments such as CIAT (Compagnia Italiana Alberghi Turistici or Italian Tourist Hotel Company), founded by Gaetano Marzotto, which in 1949 became Jolly Hotels, the first “modern” Italian hotel chain, aimed at the rising middle class. The hotel chain’s strategic positioning of its facilities has been identified as a key factor in the promotion of tourism in Italy. The presence of Jolly Hotels in specific locations has been shown to have a significant impact on the local tourism sector, contributing to the economic growth and cultural vibrancy of the areas in which they are situated.

Alongside the well-established categories of accommodation, there are alternative forms of hospitality spaces, such as campsites and holiday villages, which have been designed to cater to a new type of guest (De Falco 2021). These formats have developed from the search for forms of temporary accommodation that do not require infrastructure that alters the environment. As recognized by designer Ettore Sottsass, these new formats represent a departure from the traditional notion of vacationing as a stationary practice, giving way to a fundamentally dynamic phenomenon in which movement and travel take precedence over prolonged stays: "everyone comes to move, and everyone comes to see and be seen, to observe themselves, others, and the world in motion" (Sottsass 1953, p. 35). Temporary solutions, initially promoted by the Italian Touring Club's social holidays through campsites and lightweight facilities, later became more structured through bungalows and semi-permanent structures, finding references in the significant examples of Le Corbusier's Cabanon at Cap San Martin in 1950 and Paul Rudolph's cottage in Florida in 1951. Within a decade, the initial temporary settlements underwent a transformation into structured holiday villages, replete with numerous amenities, which have since become a common feature along many Italian coastlines (De Falco 2021).

It can therefore be posited that the entire tourism sector, and particularly hotel design, experienced significant growth and development in the postwar period and throughout the 1950s. Hotels thus became a tangible symbol of a "collective dimension of life", as emphasised by Gillo Dorfles (Aloi 1961, p. 7), as well as a showcase of Italian craftsmanship and design nascent industry. Hotels are not only a place of temporary hospitality, especially during the summer months, but also a site for the experimentation of spatial equipment solutions, furnishings, and products, as well as aesthetic languages. Hotels also become destinations themselves, with reference to specific geographical contexts that gradually established themselves, both in economically accessible contexts, such as the Romagna coast, with significant examples such as Rimini, and in much more exclusive contexts, such as Capri (Naples). While seaside destinations served as key sites for experimentation, mountain locales likewise emerged as design frontiers and destinations, as evidenced by the articles dedicated to Carlo Mollino's alpine projects, published from 1948 onward, when Gio Ponti resumed the editorial direction of *Domus* (Redazionale 1948).

3 Gio Ponti and the Spread of the "Italian Spirit" Through Hotels

In relation to a new concept of everyday life and leisure time, as well as a "positive vision" that inspired not only the design and visual arts sectors but also characterized that decade marked by a desire for rebirth, especially in Italy, Gio Ponti, first and foremost, brought a new focus to hotels and hospitality spaces. His prolific design productivity has catalysed advancements in the field of hospitality design, not only from a technical perspective, but also in terms of narrative and communicative aspects. His work has played a pivotal role in promoting and popularising an Italian lifestyle, which is widely recognised and sought after internationally. This lifestyle is exemplified in certain hotels, which have been influenced by his design philosophies. Hotels thus became, especially in their interiors, a testing ground for the construction of a language for the nascent industry that Ponti supported by combining the urgency of reconstruction to identify new models of spaces and products to meet new needs with a moment of extraordinary design creativity that dialogued with the courage and entrepreneurial capacity of small and medium-sized enterprises (Romanelli 2003). In the pre-war period, Ponti focused on the functional and technical aspects, based on the practicality and comfort of accommodation facilities, "training" himself through projects such as the Hotel San Michele, Capri Island with Bernard Rudofsky (1938), a project inspired by Italian vernacular architecture but anticipating the concept of a scattered hotel with individual units camouflaged in the local context, and projects such as the Sporthotel Paradiso in Val Martello (Bolzano). Built in 1936 and published in *Domus* in 1938, the hotel embodies some thoroughly contemporary themes, such as the hybridisation between the function of hospitality for medium- to long-term stays and the possibility of short stays.

In the period immediately after the war, Ponti continued to work extensively on the theme of hospitality. Between 1948 and 1952, he was involved in the design of naval interiors for large ocean liners such as the Andrea Doria (Rostagni 2024). The interior design of large ocean liners provided an opportunity to promote the Italian approach to design, suggesting a new way of living and experiencing travel, that *Aria d'Italia*, which gave its name to the magazine published by Gio Ponti with Daria Guarnati, whose issue 8 in 1954 was a monograph on Ponti's work (Guarnati 1954) and that can be translated as the Italian spirit (Rostagni 2024). Through experimentation with new materials obtained from the conversion of the war industry and with forms of furniture that had to fulfil new functions in relation to a more dynamic lifestyle, old furnishing patterns were overturned, and mass-produced furniture began to spread.

In 1951, Gio Ponti designed a prototype hotel room (4 × 3 m) as part of the 9th Triennale. In this design, Ponti sought to identify furnishing solutions for compact spaces that were both practical and comfortable, but also beautiful. The installation showcased a series of experiments conducted on a wall unit equipped with devices designed to execute various functions in a pragmatic manner (Scullica 1999). These included the headboard, a folding table that supported the sleeping area, an integrated lighting system, a control panel for electrical commands, a radio, and a combination of a chest of drawers and a small desk. The equipment system has been meticulously designed to be in harmony with new and functional finishing materials, including rubber flooring, which has been incorporated to ensure optimal hygiene and effective soundproofing. It is evident that Ponti plays a pivotal role in comprehending the significance of hospitality spaces in cultivating an Italian lifestyle. This notion, often referred to as the *Aria d'Italia*, a sense of the Italian cultural approach to lifestyle as well as to design (De Maio 2020) is nowadays considered the essence of the Made in Italy concept. Through his designs for hotels, Gio Ponti implemented the comprehensive approach that characterises his work and distinguishes his hospitality projects, including those by other designers, as some of the most interesting in the history of interior design.

Ponti was the first to develop the ability to connect the elements of *genius loci*, in terms of sense of place, local culture, which includes ethnographic aspects, and the conformation of the morphological context with his own imagination to define an *ante-litteram* concept (that is, a concept ahead of its time) that generated the hotel project. This ability, combined with his skill in applying every aspect of the concept to all scales of the project, in a system of elements in constant dialogue with each other, allows us to identify the origin of hotel design, as commonly understood today, in Ponti's approach to hotel design. At the same time, Ponti also considers aspects related to art, decoration, graphics, and communication, according to a coherent and unified vision that manages to combine the organisational elements of spatial layout and interior furnishings with decorative and artistic elements in a way that is synergistic with narrative and communicative elements, which today we might define as an expression of brand identity. Gio Ponti thus outlines a narrative path that connects the individual elements with the environments that become true experiential stages for guests, who are "on stage" and "protagonists" in a way that anticipates the concept of the hotel as a "stage". This concept would revolutionise hotel formats with the boutique and design hotels of the 1980s and 1990s, from the creations of Jan Schrager, a New York entrepreneur who moved from the entertainment sector to the hotel industry, calling on emerging designers such as Philippe Starck to reinvent the format, onwards (Scullica 1999).

Gio Ponti's most significant hospitality design projects are concentrated in the Naples area, where Ponti encountered a particular conception of Mediterranean architecture. In the Hotel Royal project in Naples (1953), the architect was tasked with designing the interiors and transforming the rooftop into a habitable space by constructing a swimming pool and an outdoor social area to cater to the city's jet set and tourists. The swimming pool was designed in the shape of an irregular polygon broken by a constriction and was entirely covered with colored ceramics, as were the surrounding spaces. This approach had already been experimented with by Ponti in 1948 with the swimming pool of the Hotel Royal in San Remo, achieving a very communicative result. Furthermore, Ponti was responsible for establishing the coordinated image of the structure, designing the logo-coat of arms, and choosing a color scheme to guide the furnishings. Subsequently, Ponti was commissioned by the same client to design two Hotel Parco dei Principi projects in Sorrento and Rome, which are widely regarded as the most paradigmatic examples of the architect's approach to hospitality design.

4 The Hotel Parco dei Principi in Sorrento, 1960–1961

The design of the two Hotels Parco dei Principi is a paradigmatic example of Ponti's design approach. Two projects, thematically analogous, were commissioned by the same client, the Neapolitan entrepreneur Roberto Fernandes (Mautone 2009).

The hotels were situated in two dissimilar contexts: the Sorrento coast for the Hotel Parco dei Principi, located in the park of Villa Poggio Siracusa, Sorrento

(Naples), designed and realized in 1960–1961, and the city of Rome, close to Villa Borghese, for the Hotel Parco dei Principi, designed and realized in 1961–1964. The conception of both hotels occurred in the early 1960s; however, their design and ethos are significantly influenced by the 1950s (Scullica 1999). Despite their different locations, the two hotels are united by their use of color. In the Sorrento project, the color scheme is based on blue and white, while in Rome it is green and white. The decoration of the hotels also strongly accentuates the theme of the indoor-outdoor experiences.

However, the Sorrento Coast's distinctive topography and landscape, in comparison to the notable Roman influences in Rome, will ensure the first Parco dei Principi hotel's status as an iconic destination. The hotel "overlooking the sea" will become an iconic destination, and it has maintained its popularity and has remained largely unaltered to the present day, while the Roman hotel has undergone renovations over the years that have significantly altered the original design. Parco dei Principi Hotel in Sorrento serves as a prime example of the Italian lifestyle expression in hospitality, because its atmosphere, created by the skilful relationship between landscape and architecture, between interior and exterior, and the almost obsessive attention to every detail,[2] is still highly appreciated by international tourists today.

The Hotel Parco dei Principi in Sorrento is an extraordinary project that embodies the traditional values of Italian design, setting the standard for hospitality design.

The Italian approach to hospitality design is epitomised by:

1. The relationship with the place, understood in terms of its territorial, morphological, and ethnographic aspects. Ponti was commissioned to work on an existing building within the grounds of Villa Poggio Siracusa, an area with significant historical features.
2. The relationship between craftsmanship and industry. In this project, Ponti combines craftsmanship for the creation of certain surface finishes and decorative elements with production processes aimed at mass production, as in the case of D'Agostino ceramics and Ceramica Joo pebbles (La Pietra 1995). The same approach applies to industrially produced furniture systems by emerging brands, such as Cassina. Ponti's research, based on experiments and studies in the decorative arts to extract those characteristics conducive to mass production, is very significant in the Parco dei Principi project (Mautone 2009). Among the examples that can be identified are the 20 × 20 cm ceramic tiles made by Ceramica D'Agostino, in white and blue, with 27 different decorations personally designed by Ponti, allowing for infinite compositions. These ceramics constitute the module, simple but firmly connected to the concept and the territory of telling a hundred different stories, each enclosed within one of the hotel's hundred rooms. As Ponti explains about the Sorrento project, "give someone a 20 × 20 square and, although over the centuries everyone has indulged in endless designs, there is always room for a new design, for your design, there will never be a final

[2] For a discussion on unconventional places where Made in Italy can be found, see Bollati and Collina's essay in this volume.

design" (Ponti 1964, p. 39). The industrial element is conceived here as a significant component of the project, with its own aesthetic dimension, which allows for the variety and complexity that the designer seeks while keeping costs and construction times down.
3. The role of art. Collaborations with artists, particularly with regard to the ceramics created by artist Fausto Melotti, complete Ponti's narrative journey, which integrates all the elements that complete the project into the individual rooms.

These components enable the creation of a "total design" for rooms, combining spaces, equipment, products, finishes, and decorations with the lifestyle of the era—often linked to a specific location—to produce a coordinated project.

Ponti's approach as a contemporary art director can be seen in his obsessive attention to detail, as well as in the design of the experience. One interesting anecdote recount how Ponti invited the hotel chef to use methylene blue when preparing the dishes to be served at the opening, ensuring that the food would fit in with the overall design.

The colors blue and white, symbolic expressions of the sea and sky, and the (non-)color white, symbolic expression of the dazzling Mediterranean light, are key elements of the concept, together with the decoration and artistic dimension. Together, they form a journey for guests, leading them from the outside to the inside, and from the communal areas to the rooms and restaurant. This rhythmic sequence of spaces is designed according to specific visual perspectives, which Ponti referred to as "visual eyes". In his drawings of plans for domestic spaces such as the Villas Planchart and Arreaza (in Caracas, Venezuela), and Villa Nemazee (in Teheran, Iran), as well as the ground floor plan of the Parco dei Principi, he used to trace "optical cones" (Mautone 2009, p. 91) to study the point of views and the sequences as a filmmaker. This approach means that each functional space constitutes a separate episode within a stylistically coherent vision and narrative path. Each of Ponti's choices is not merely formal; they also respond to the functional and service needs of an international clientele with whom Italy is reconnecting after the "autarchic" closures of the previous period. The park and its residential buildings, which belonged to Princess Elena Cortchacow (a relative of the Russian tsars), inspired the design of the hotel, as did the colors and light. In an article in *Domus* 415 in 1964, Ponti emphasises these elements in the title "Blue sky, blue sea, blue islands, blue majolica, green plants, roses at the princess's feet, a dancer's footprint" (Ponti 1964, p. 39). These chromatic elements became an almost obsessive inspiration, as did the reference to the prima ballerina at the Court of St. Petersburg, who performed in the residence during its years of splendour and vitality. The villa's history and the mysterious figures who stayed there add an element of "glamour", generating further interest and attention for guests seeking the splendour and most significant elements of the past, but with a contemporary interpretation. The park features an irregularly shaped swimming pool, which has already been tried and tested in the Hotel Royal in Naples (Fig. 1). In this context, it echoes the location's rocky ruggedness and influences the building's overall architectural composition, with a diving board that resembles a neo-expressionist work of art from an art gallery. This new approach is

Fig. 1 The outdoor swimming pool with an organic shape. (Image courtesy of Hotel Parco dei Principi Sorrento)

in line with the aesthetic preferences of an elite clientele who are attentive to both contemporary fashion and their own lifestyle. Furthermore, the swimming pool's generous dimensions bring guests closer to the sea, which is not particularly accessible as it lies at the foot of the high cliff on which the hotel stands.

The overall architecture interacts with the morphology of the site through masterfully articulated, composite forms that echo the verticality of the pre-existing architecture on which the new hotel structure is built. As far as the interiors are concerned, despite what one might initially think, Ponti does not take an exclusively formal and stylistic approach in this project, as he does in many others. Rather, he takes the approach of a designer who skilfully blends the various components that form the basis of every project, such as functionality, technique, aesthetics, and communication, in order to design an unforgettable experience for every guest (Scullica 1999). The various components are integrated at all levels of architecture, interior design, and art. This is particularly evident in the solutions adopted for horizontal coverings and vertical elements such as walls and pillars, particularly with regard to research into ceramics. These elements are covered by a background texture of majolica elements (Mautone 2009) which were created in collaboration with Ponti's friend, the sculptor Fausto Melotti. They are directly related to the areas covered with ceramic pebbles by Joo, emphasising the continuous dialogue between art, craftsmanship and industry that characterises the hotel's interiors (Figs. 2 and 3).

All the rooms are identically furnished with Cassina furniture, but each has different patterned floors created from a composition of strictly white and blue

Figs. 2 and 3 (above) and (below) In common living spaces on the ground floor, the patterns and the materials expressions of horizontal and vertical surfaces (decorated tiles, majolica elements, and ceramic pebbles) are combined. (Image courtesy of Hotel Parco dei Principi Sorrento)

ceramics. This reflects the extreme functional and technical rigour of the room and furnishing design. One of the cornerstones of Ponti's furniture designs, the "dashboard furniture", already tried and tested in many other interiors such as his home in Via Dezza in Milan and the Hotel Royal in Naples, finds further significant application and variation in the Hotel Parco dei Principi. From this point of view too, Gio Ponti is the precursor of aesthetic research, explored through multi-material furnishings in ash, Formica laminate, and metal elements with a matt nickel finish. In fact, "every piece of furniture, while always being functional (the functions of furniture are various and include pleasure), should move the imagination of those who design it and those who look at it" (Mautone 2009, p. 152).

Gio Ponti anticipates the extreme customisation in terms of variety, size, and finishes that characterises today's contract production for the hospitality sector. The experiments, which were highly innovative and research-oriented for the time, were carried out in the design of common areas, but above all in the furnishing of bedrooms, and constitute the fundamental references for contemporary interior design for hospitality. Ponti goes beyond the systemic approach to furnishing, focusing on individual pieces that interact with each other in the bedroom but can also be used separately in other contexts.

In fact, the functionality of the solutions adopted still makes this furniture valid today for meeting the needs of customers. Currently, the hotel rooms still feature bed headboards equipped with push-button panels for lighting the room and sleeping area, as well as easily accessible shelves to ensure the comfort and needs of the "modern" guests of the time (Fig. 4).

Armchairs and other furnishings are designed with expressive shapes but are compact in size to leave room for the floor decorations and the view of the outside

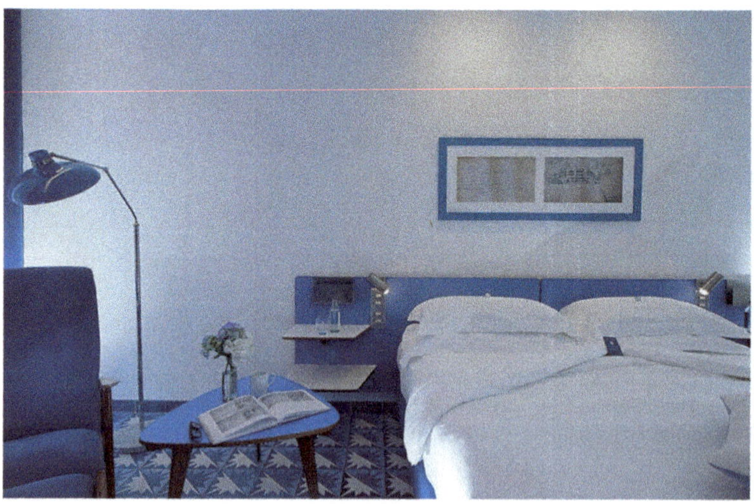

Fig. 4 All the hotel rooms are characterised by personalized floor decoration, equipped bed headboard and selected furniture. (Image courtesy of Hotel Parco dei Principi Sorrento)

landscape, in line with the tradition of historic Mediterranean houses, rich in decorations and colors in the finishes of the surfaces (floors, walls, and ceilings, etc.) but more essential in terms of furnishings. The focus on "tradition" is evident in the redesign of certain objects and furnishings, including a reinterpreted dressing table (Scullica 1999).

These welcoming spaces, therefore, lend themselves to experimentation and testing of products such as the iconic Superleggera chair. Although not specifically designed for the hotel, it is available in the project's finishes and is widely used in the rooms due to its lightness, which makes it easy to move during cleaning or when in use. The "slenderness" of the furnishings ensures ample freedom to move them around without affecting the beauty and order of the rooms (Mautone 2009).

The room furnishing system recognises the promotional value of hotel spaces as an opportunity to introduce guests to desirable objects and accessories that can be purchased, anticipating marketing strategies that characterize hospitality design today.

From private spaces to the restaurant, the project not only defines the common areas for socialising and leisure and their furnishings, all produced by Cassina, but also addresses rituals such as meals and breaks, which are becoming increasingly important in the new lifestyle.

Ponti also focuses on the definition of the furnishings for the dining areas, designing the *mise en place* and many of the complementary items and accessories that comprise it, such as textiles, but also cutlery, plates, and glasses, right up to the level of "food design" with the design of the blue dishes mentioned above.

5 Conclusion

Today, hotels such as the Parco dei Principi remain icons, bridging past and present by preserving their legacy while continuing to embody the enduring timelessness of Italian design. Indeed, the design approach adopted for the Hotel Parco dei Principi is emblematic not only of Gio Ponti's "style", which was subsequently applied to numerous other projects, but also serves to recognise the hotel as the ideal setting for expressing an all-Italian lifestyle that exerted a significant influence on the definition of Made in Italy throughout the 1950s. Furthermore, it provides a valuable insight into the approach to hospitality design taken by subsequent Italian designers. Through this project, Ponti anticipates the brand's declination and governance strategies, which today are reflected in the missions of large hotel chains as well as in individual projects and in the most advanced brand extension or co-branding operations, which currently characterise some of the most significant developments in high-end hospitality. Projects that have gained global recognition as emblematic of contemporary Made in Italy, such as Bulgari Hotels, are founded on the design tradition and legacies established by Ponti.

The ACPV ARCHITECTS Antonio Citterio Patricia Viel studio has adopted Ponti's approach to the overall design of Bulgari Hotels, translating the essence of the Italian jewellery brand, which is strongly connected with Italian identity and culture, into different systems of products, spaces, and services (Scullica and Elgani 2025; Cervetta and Scullica 2007). The meticulous research and interpretation of the brand's essence, particularly its quintessence of "Italianness" (Dallabona 2019) and a distinct Italian *savoir-faire*, which draws from craftsmanship and the symbiotic relationship between craftsmen and industry, render the Italian spirit espoused by Gio Ponti immediately discernible (Colonnetti and Famiglietti 2017).

Competing Interests The authors have no conflicts of interest to declare that are relevant to the content of this chapter.

References

Aloi G (1961) Alberghi. Motel. Ristoranti. Hoepli, Milan

Battilani P (2001) Vacanze di pochi, vacanze di tutti: l'evoluzione del turismo europeo. Il Mulino, Bologna

Battilani P (2020) Gli anni in cui tutto cambiò: il turismo italiano fra il 1936 e il 1957. Transportes, servicios y telecomunicaciones 41(1):103–133

Battilani P, Fauri F (2020) Marshall plan help to the airline sector and its impact on the development of tourism in the Italian regions. In: Knoll M et al (eds) Transformative recovery? The European Recovery Program (ERP)/Marshall plan in European Tourism. Innsbruck University Press, Innsbruck, pp 91–116

Berrino A (2021) Visioni e ruoli del Touring Club Italiano nel dibattito e nel riassetto del turismo nel secondo dopoguerra. In: Storia dell'urbanistica. Annuario nazionale di storia della città e del territorio, vol 1. Caracol, Palermo, pp 149–165

Cervetta G, Scullica F (2007) Il Bulgari Hotel. Caso studio di integrazione fra design, moda, servizi. In: Bertola P, Conti G (eds) La Moda e il Design. Il trasferimento di conoscenza a servizio dell'innovazione. Edizioni POLI.design, Milan, pp 159–166

Colonnetti A, Famiglietti P (eds) (2017) Gio Ponti: L'infinito Blu. Edizioni Paguro, Mercato San Severino

Corbin A (ed) (1995) L'avènement des loisirs, 1850–1960. Aubier, Paris

Dallabona A (2019) Be Italian! Aesthetic employees and the embodiment of national traits in luxury fashion flagship hotels. In: 2019 IFFTI conference. Manchester Fashion Institute. https://fashioninstitute.mmu.ac.uk/assets/uploads/2019/07/59-Be-Italian-Aesthetic-employees-and-the-embodiment-of-national-traits-in-luxury-fashion-flagship-hotels-Alice-Dallabona.pdf. Accessed 1 Jul 2025

De Falco C (2019) Notizie dal mondo: alberghi e turismo nelle riviste del Touring Club Italiano. In: Belli G, Castagnaro A (eds) Le città e il turismo. Hotel tra Ottocento e Novecento. Arte'm, Napoli, pp 27–39

De Falco C (2021) Il Touring Club Italiano e la promozione della vacanza nei villaggi negli anni Cinquanta, dalla montagna al mare. In: Storia dell'urbanistica. Annuario nazionale di storia della città e del territorio, vol 1. Caracol, Palermo, pp 235–250

De Maio F (2020) 1 villa, 3 alberghi. Le occasioni di Gio Ponti a Napoli. La Rivista di Engramma 175:53–73

Deambrosis F, De Magistris A (2015) Tempo libero e progetto. Prospettive italiane tra gli anni '50 e '60. Territorio 83:114–120

Guarnati D (1954) Espressione di Gio Ponti. Aria di Italia VIII

Kawamura E (2021) Repertorio e tendenza negli articoli della Rivista del Touring Club Italiano "L'albergo in Italia" (1925–1943). In: Storia dell'urbanistica. Annuario nazionale di storia della città e del territorio, vol 1. Caracol, Palermo, pp 275–298

La Pietra U (1995) Gio Ponti. L'arte si innamora dell'industria. Rizzoli, Milan

Le Corbusier (1923) Vers une Architecture. Les Éditions G. Crès, Paris

Mautone F (2009) Gio Ponti e la committenza Fernandes. Electa, Naples

Pasquali A (1940) Introduzione alla mostra sull'arredamento alberghiero. In: Catalogo VII Triennale. Triennale di Milano, Milan

Penner A, Rutes W (1985) Hotel planning and design. Architectural Press, New York

Ponti G (1948) Che cosa può significare la parola "funzionale": discorso su una piscina. Domus 229:10–13

Ponti G (1964) Ritorno al Parco dei Principi. Domus 415:39

Redazionale (1948) Tre architetture in montagna: costruzioni e progetti dell'architetto Carlo Mollino. Domus 226:9–21

Romanelli M (2003) Gio Ponti: a world. Abitare Segesta, Milan

Rostagni C (2024) "Fantasia degli italiani". Il Made in Italy secondo Gio Ponti. In: D'Acunto G et al (eds) Prospettive del Made in Italy. Libria, Melfi, pp 171–179

Scullica F (1999) Lo spazio privato alberghiero. Realizzazioni e osservazioni disciplinari. Dissertation, Politecnico di Milano

Scullica F, Elgani E (2025) Fashion brand extension processes in hospitality. The Bulgari Hotels and Resorts case study and the sustainable design strategies for future development. Paper presented at Designing Retail & Services Futures Colloquium, Design Research Society, Politecnico di Milano, 14–15 May 2025

Sintini M (2018) Gardella/l'abitare: lo spazio temporaneo. In: FAMagazine. Ricerche e Progetti sull'architettura e la città 44: 22–28. https://doi.org/10.12838/fam/issn20390491/n442018/148

Sottsass E Jr (1953) Alberghi e tende. Domus 287:34–36

Archives for the Memory and for the Future

Design Archives, Made in Italy, and Communication in CASVA's Recent Activities

Maria Teresa Feraboli

Abstract This essay explores communication strategies "Made in Italy" through the recent activities of the CASVA (Centro di Alti Studi sulle Arti Visive, or Centre for Advanced Studies in the Visual Arts), focusing on its design archives. The CASVA has been using modern tools, mainly social media (Facebook), to broaden its outreach and bring greater visibility to such iconic Italian designers as Luciano Baldessari, Enrico Freyrie, Augusto Magnaghi, Mario Terzaghi, and Roberto Sambonet. By curating digital content from its vast design archives, the CASVA has successfully let Italy's rich design heritage reach a broader, global audience. Through social media platforms, it has created engaging virtual exhibitions and interactive features that celebrate Italian craftsmanship and make it accessible to contemporary audiences. This strategic use of digital communication marks a significant effort to democratize cultural heritage, ensuring that the "Made in Italy" legacy reaches a diverse, far-reaching audience.

Keywords Design archives · Made in Italy · Roberto Sambonet · Enrico Freyrie · Augusto Magnaghi · Mario Terzaghi

1 Introduction

The CASVA (Centro di Alti Studi sulle Arti Visive, or Centre for Advanced Studies in the Visual Arts) is a cultural institute of the Municipality of Milan, operational since 2002, and tasked with safeguarding sources for the study of visual arts in the Milan area. Step by step, and thanks to the first major acquisition (a group of sketches, drawings, and paintings by the architect-artist Luciano Baldessari), the institute began to gather and preserve material and make its archives of Milanese and Lombard designers available both to scholars and a broader non-specialist

M. T. Feraboli (✉)
Politecnico di Milano, Milan, Italy
e-mail: mariateresa.feraboli@polimi.it

P. Cordera (ed.), *Unfolding Made in Italy (1948-1962)*, Research for Development, https://doi.org/10.1007/978-3-032-10118-1_4

audience. In 2014, thanks to the direction of Dr. Maria Fratelli,[1] the CASVA made its mission clearer by adding a meaningful subtitle to the name, i.e., *The Design Archives* (Fratelli 2021), thus including the results of a wide-ranging creativity which encompasses technical-design aspects on top of purely artistic ones. Indeed, there are several collections from architects or designers who also successfully frequented the art world: by way of example, and in addition to the aforementioned Luciano Baldessari, Roberto Sambonet, Enzo Mari, and the latest acquisition, the Nanda Vigo archive.[2] Among the more recent acquisitions are those of the artist-designer Eugenio Carmi and Alfonso Pizzo Greco, which help strengthen the multidisciplinary targets of the institution's origins and breadth of interests. The CASVA now possesses a patrimony consisting of 45 archives[3] and has finally solved the long-standing problem of its temporary location split between the Sforza Castle and the Fabbrica del Vapore arts venue, with this year seeing the opening of a new headquarters at the former indoor market in QT8,[4] a symbolic district of Milanese modernity, born under the aegis of the 8th Milan Triennale (1947).

2 CASVA's Communication Strategy

CASVA's communication takes place through channels and methods designed to share the initiatives, events and acquisitions of the centre both in an institutional way (official website, press releases, newsletters, scientific publications, exhibitions, etc.), and in a freer and more informal way (social media, promotional novelties, etc.), but always based on an in-depth look at the archives and collections in its

[1] Maria Fratelli has been director of CASVA since 2012.

[2] On alternative approaches to designers' archives, see Chiesa and Scalco's essay in this volume.

[3] The CASVA collections made available for this study totalled 21: those of Béla Angelus, Cecilio Arpesani, Luciano Baldessari, Giancesare Battaini, Angelo Cortesi, Marcello Cuneo, De Pas, D'Urbino, Lomazzi, Andrea Disertori, Fredi Drugman, Enrico Freyrie, Ginoulhiac-Arslan, Francesco Gnecchi Ruscone, Vittorio Gregotti (Gregotti, Meneghetti, Stoppino and Gregotti Associati studios), Amneris Latis, Augusto Magnaghi and Mario Terzaghi, Enzo Mari, MID Design, Fiorenzo Ramponi, Roberto Sambonet, Eugenio and Ermenegildo Soncini, and Virgilio Vercelloni. Not yet available for consultation, but acquired since the Covid pandemic years, are the archives of Matilde Baffa and Ugo Rivolta, Jan Andrea Battistoni, Eugenio Carmi, Antonio Cassi Ramelli, Marcello Cuneo, Eleonora De Benedetti, Vittorio Faglia, Eleonora Fiorani, Angelo Galesio, Alberto Grimoldi, Manifattura JSA, Hiromichi Matsui, Zita Mosca Baldessari, Egizio Nichelli, Anty Pansera, Maria Pezzi, Franca Parisi Baslini, Walter Pastcheider, Alfonso Pizzo Greco, Mario Salvadé, Eugenio Fabio Simion, Nanda Vigo and complementing Cecilio Arpesani's collection, the Cattoretti villa project, donated by the Cattoretti family.

[4] The project to renovate the former indoor market was born in 2017, starting from the studies of a working group coordinated by Mariella Brenna (DASTU—Department of Architecture and Urban Studies, Politecnico di Milano) with the aim of transforming the building into a space for the conservation, consultation and exhibition of the CASVA archives. The project, brought to a conclusion by Walter Patscheider and Corrado Serafini of Architects for Milan workshop and entrusted to the MM company, became a worksite in the autumn of 2021.

possession. This strategy blends concrete physical activities with digital or virtual initiatives, as envisaged since the first studies by Jeffrey Schnapp on the evolution of archives and libraries in the era of Digital Humanities (Schnapp 2013; Schnapp and Mattei 2015).[5]

Of fundamental importance are the ever-new "narratives" that arise from analyses of documents, drawings, prototypes, and models presented in traditional ways (printed publications, brochures, celebratory postcards, exhibitions) as well as more technologically advanced ways, adapted to the era and to today's interlocutors (online exhibitions, social media images and comments). In fact, the CASVA carries out a non-stop parallel "re-mediation" of archival-humanistic skills using contemporary tools, adopting languages and techniques that can "mediate" between the past preserved in its depositories, the present, and the future (Ferrario 2024). A continuous exchange between analogue and digital communication is pursued, often feeding one with the other's developments. One example of this is the homepage of the institutional website[6] which uses a series of postcards[7] as scrolling announcements which, in 2015, celebrated the inventoried archives of the CASVA, in parallel with the exhibition *Milano Città Immaginata—Milan, a City Imagined*, created in an analogue version with a temporary duration, as well as online, which remains consultable;[8] the postcards and exhibition as a whole brought home the importance of providing the institution with a definitive location.[9] In addition, *Milano Città Immaginata* was advertised by ten posters (one for each project on display), entrusted to a group of Milanese illustrators,[10] who freely reinterpreted the originals on show (Sacerdoti 2020; CASVA 2015). Confirming the close relationship between physical and virtual activities, it is worth recalling that the exhibition was followed by a catalogue (CASVA 2015) and a hefty volume (Sacerdoti 2020), as well as a header on the CASVA Facebook page. The latter was a further reworking of the posters designed to promote the exhibition, starting from a visionary project by Jonathan De Pas, Donato D'Urbino, and Paolo Lomazzi with Roberto Sambonet for the Cathedral Square in Milan (1984), kept at the institute (Fig. 1).

---

[5] On the application of Schnapp's principles in practice, see Costa and Di Pietro's essay in this volume.

[6] The official website, supported by that of the Sforza Castle in Milan, is currently being reorganized.

[7] Designed by Enrico Delitala, the postcards were dedicated to the archives of Luciano Baldessari, De Pas D'Urbino Lomazzi, Vittorio Gregotti, Roberto Sambonet, and Andrea Disertori. The online version features altered proportions.

[8] *Milano Città Immaginata* was held at the Guido Nardi exhibition space of the Politecnico di Milano from 16 to 23 October 2015. Now available via https://milanocittaimmaginata.it/it/index.html. Accessed 20 August 2025.

[9] The exhibition, held at the Politecnico di Milano, was a tribute to the culture of Milanese design and, at the same time, an open reminder of the need to safeguard the documents that testify to this culture, flanked by a series of conferences culminating in the *Cerco CASVA* round table at the former QT8 market designed by Piero Bottoni, and currently the CASVA headquarters.

[10] The members of the working group, coordinated by Enrico Delitala, were Luca D'Urbino, Lorenzo Gaetani, Camillo Lentini, and Anna Sutor.

Fig. 1 Poster of the exhibition *Milano Città Immaginata* (2015) on the Facebook CASVA Homepage

During the horrendous COVID-2019 pandemic, social media and Facebook in particular proved useful participatory tools: they supported the institute's visibility and, above all, helped maintain and strengthen existing bonds with donors, collaborators (inventors, interns, photographers, graphic designers, etc.), scholars, and simple fans of architecture, design, and art. This experience, set in motion by Elisabetta Pernich and now being continued by Virginia Invernizzi, has made it possible to increase that sense of a participatory community which the director, Maria Fratelli, has been able to create around the CASVA. A feeling also corroborated by the *Highlights* to be found on the institutional website, i.e., descriptions of the work of designers who can be reached through their photographic portraits,[11] which, wherever possible, were taken during public events organized by the CASVA itself: a tangible demonstration of the direct involvement of donors who have become an integral part of the institute's life.

[11] These are photographic portraits of Francesco Gnecchi Ruscone and Giances are Battaini; in other cases, the *Highlights* feature emblematic products, works of architecture, or drawings.

3 "Stories of Made in Italy (1948–1962)" in CASVA's Communications

The most recent studies attribute to the expression "Made in Italy" a broad meaning which goes beyond the exact geographical origin and expresses the best characteristics of Italian creativity and artisan and industrial production, strengthened by the indissoluble relationships with the cultural, architectural, and environmental contexts that characterize Italy's uniqueness in the world. The post-World War II period, with the tragic nature of the immediate reconstruction behind it, became the forge of the country's social, economic, and productive rebirth, which, helped by the export of products—especially to the United States—and the mounting of exhibitions, made everyday objects synonymous with a much-admired lifestyle (Dellapiana 2022).[12] Using the Made in Italy term today is almost equivalent to talking about a "brand", which, like a delicate fabric, is born from an interweaving of multiple threads that determine its warp and weft. These same threads also snake through the CASVA archives and connect them, creating numerous possible stories which, during the years of the pandemic, Elisabetta Pernich began to sketch concisely and scientifically in Facebook posts, creating special albums in which many of the elements that nourished the beginnings of the Made in Italy concept can be picked out.

These particular albums, which remain active, range from portraits of designers to Compasso d'Oro awards, from in-depth analyses of individual products to layouts for exhibitions at the Milan Triennale, from works of architecture to direct testimonies of protagonists or heirs.[13] These posts were conceived to inspire both in-depth and generic study, "delicious appetizers" of exquisite dishes stored in the institution's depositories. During the COVID pandemic, the main objective was to keep interest in CASVA materials alive without any possibility of physical visits. This was combined with the intention of sparking dialogue among followers while reaching out to other archival institutions through specific links on common topics, reactivating interrupted relationships:[14] in fact, the CASVA had always participated in the National Archive Day, established by the AAA Italia (Associazione Nazionale Archivi Architettura, or Italian Association of Architectural Archives) in 2011, and in the Festival and Night of the Archives of Archivissima, born in 2017.

The collections possessed by the CASVA[15] testify to the conception, design, and production of brand-new, functional, and beautiful household objects, suitable for a society and a lifestyle that were reborn and transformed between 1946 and 1960. At the same time, they also attest to the promotion of this renewal thanks to

[12] On exhibitions in unconventional places in a historical perspective, see Cordera's essay in this volume.

[13] The albums have titles like *Archives tell their stories, Compasso d'Oro, Triennials*, but also *Projects on the Road, Summer*, etc.

[14] Thanks to the posts, the followers increased from around 300 to 1910 on Facebook and 722 on Instagram.

[15] The study refers to the collections and posts present at the date of the conference.

Fig. 2 From the Facebook album *Compasso d'Oro*: the fish kettle (1957), awarded in 1970

commercial and cultural installations at the Milan Trade Fair, la Rinascente department stores, and the Milan Triennale, against a backdrop of the city's reconstruction and renewed growth. It is, above all, the posts in the *Compasso d'Oro*, *Allestimenti* and *Gli Archivi si Raccontano* albums, intelligent monographic gleanings dedicated to both works and protagonists (in particular Roberto Sambonet, Amneris Latis, Enrico Freyrie and Augusto Magnaghi with Mario Terzaghi), which outline the appearance of works that are truly Made in Italy in the CASVA's collections. It is also worth emphasizing that the posts of the various albums contain clear references to each other by means of digital cross-references, but also to other archives: a demonstration of how it is an inextricable interweaving of various stories that has formed the basis of the ideal, critical, and commercial fortune of the Made in Italy concept.

On examining the *Compasso d'Oro* album,[16] it can be seen that it holds 12 CASVA archives, from the origins of the award to the present day; in the period of interest to this paper, the posts show in sequence some of the winners of the first editions since the establishment of this award. These include the SAFFA modular kitchen by Augusto Magnaghi (1954), the Idrosci Universal water skis by Enrico Freyrie for Fratelli Freyrie (known as Freyrie Ski since 1955), and the round-oval centrepiece plate of the stainless steel tray series by Roberto Sambonet for the Sambonet company (1956); later (in 1970), the artist designer repeated his success with his famous fish kettle of 1957, again for his family business.[17] These are all products which testify to the building and achievement of a lifestyle that speaks of modernity, elegance and wellbeing, in particular the fish kettle (Fig. 2): the domestic

[16] The Compasso d'Oro, the top Italian award for design quality, was established in 1954 from an idea of architect and designer Gio Ponti and launched by la Rinascente company; since 1966 it has been run by the ADI (Associazione per il Disegno Industriale, or Association for Industrial Design).

[17] Sambonet won the Compasso d'Oro award four times: in 1956 for a series of stainless-steel trays, in 1970 for the fish kettle, in 1979 for the symbol of the Lombardy Region (with Bruno Munari, Bob Noorda and Pino Tovaglia) and in 1995 for his career.

dream of the bourgeois lady thanks to being both elegant and functional, as a serving plate or alternatively a table centrepiece, a clear indication of a contemporary and refined lifestyle, as suggested in the photos by Aldo Ballo or Serge Libiszewski.

Using cross-referencing in examining the albums, the posts reveal that the authors who were awarded the Compasso d'Oro—Magnaghi, Freyrie, and especially Sambonet—had radically different careers (Carones and Landsberger 2006; Feraboli 2006; Iannello 2016) but were united by product design: objects for domestic use by Sambonet, sports by Freyrie, furnishings by Magnaghi (also in collaboration with his business partner Mario Terzaghi). From the posts emerges the well-known ability of companies, often still at the level of mechanized craftsmanship, to open up to designers' creativity, thus heading unrelentingly towards a quality industrialization that helped affirm the Made in Italy concept. Hence the post dedicated to the Bilux sofa (1955), created by Fratelli Busnelli (Meda) to a design by Magnaghi and Terzaghi for the building contractor Guffanti who, after seeing Marco Zanuso's Sleep-O-Matic, asked the architects for a sofa bed for his own home (Fig. 3). Bilux allowed the company to set up a primitive assembly line, and the sofa was offered for sale by la Rinascente and the other four European department stores, of which it was the prime contractor, thereby increasing the spread of Italian-made products abroad (Pansera and Abis 1998). Although little known, Bilux also demonstrates how many Milanese designers worked alongside the most famous names with intelligence and skill, transforming not only the architecture of "Modern Milan", but also products and interiors, resulting in an extension of the field of action of "cultured professionalism" (Irace 1996, 2021).[18]

A supporter of the birth of design from the foundations of an evolving craftsmanship towards an industry that never neglected quality was Enrico Freyrie, in a video included in the *Triennali* album. Recalling the mounting of an exhibition on *Form and Colour in Sport* at the 9th Triennale in 1951, created with Vittoriano Viganò and awarded a Gold Medal,[19] he said: "these forms of sport that were, for example, skeletons of motorcycles, were a bobsleigh, skis, or saddles—already beautifully designed. I remember a racing skiff, made from sapele wood, of exceptional lines; so I would say that in Italy, in '51, there was already unconsciously a whole series of artisans or industries working with an extremely refined design sense."[20] The exhibition also featured a Piaggio Vespa, a Guzzi twin-cylinder motorcycle and a Gilera four-cylinder motorcycle, a Bianchi racing bicycle, a sailing boat by Ernesto Zardo, skiffs by the Donoratico Shipyards, H49 and Condor skis by the Freyrie

[18] "Cultured professionalism" was a critical category developed by Fulvio Irace to indicate those designers who, in the years after the Second World War, contributed to a quality urban transformation alongside the better-known names. Today, their commitment is recognized as extending from architecture to interior and product design.

[19] The designers also enjoyed the collaboration of the sculptor Umberto Milani and the painters Roberto Crippa and Gianni Dova, all three authors of interventions placed in the rooms laid out by Luciano Baldessari.

[20] These appear at 1.14–1.46 min.

Fig. 3 From the monographic album dedicated to designers: *Augusto Magnaghi and Mario Terzaghi and the design of the '50s*: the Bilux sofa, Busnelli, c.1955. Comune di Milano—CASVA ©Photograph by Aldo Ballo

Brothers,[21] Kartell ski racks, Moretti camping tents, etc. (Pica 1951), asserting the refinement of a mass production which never forsook attention to detail, but also the key role of exhibitions in promoting products to the public. Moreover, shortly thereafter, the Vespa scooter would go on to worldwide fame thanks to the film *Vacanze Romane*, or Roma Holidays (1953), starring Audrey Hepburn and Gregory Peck against the wonderful backdrop of Rome: the cinematographic atmosphere and, above all, the image of the city contributed to internationally enhancing the cultural significance linked to the product itself. Moreover, in this regard, the skis (for both water and snow) designed by Enrico Freyrie for the family business should not be forgotten: they would quite literally become frontrunners, not only thanks to the Compasso d'Oro award. They were chosen and used by various Italian and foreign sports champions, achieving numerous victories. Brochures, advertising leaflets, and magazines exalted Italy's technical and creative primacy, combining the

[21] The company was founded in 1927 by the brothers Emilio (father of Enrico Freyrie) and Gian Giacomo Freyrie who, thanks to their passion for sport, bought and converted a carpentry shop and dedicated themselves to the production of equipment for leisure, above all skiing. They became highly successful, and, in 1947, Enrico began to collaborate by immediately introducing the use of colour in skis: the H49 model being the first of this new series.

achievements of sport with the devices that had allowed them, amplifying the success of Made in Italy products.

Returning to the *Triennali* album, Roberto Sambonet's works for the company of the same name[22] are also often present among the *Produzioni* (or Productions) dedicated to metals: in addition to the aforementioned trays and the fish kettle, in 1960, the designer exhibited the more pragmatic Electrogas pots, awarded a Gold Medal. This set, with its more traditional image than other works, was judged by the critic Arturo Carlo Quintavalle as being in line with the functional rigour and harmony of volumes that marked the designer's style, expressing "the strength of a system of culture" and "of a civilization that goes beyond the system of objects" (Quintavalle 1993). It almost seems that the scholar had summarized ahead of time the qualities that we attribute today to the Made in Italy concept.

Then there are other posts from which it can be seen how this concept resulted from a constant work of creation and promotion, also thanks to commercial exhibitions dedicated to the commitment of Amneris Latis as art director of la Rinascente company. A modern-day independent woman and a traveller, Latis led the department store chain in the 1950s, years in which Roberto Sambonet collaborated as both a graphic designer and a window designer (Fig. 4). Refined exhibitions were also set up for sales, in a position to educate the taste of a broader and less sophisticated audience than that of the Triennale, but with equal commitment and just as much ability to communicate an idea of living and lifestyle (Irace 2017). The very institution of the Compasso d'Oro at the department store is confirmation of this, and the posts, intertwining the stories of Sambonet and Latis, tell it, portraying shop windows, clothes, objects, posters, promotional items, and advertisements. It is the latter (but not only) which often animate CASVA's communication beyond Made in Italy and historical in-depth analyses: albums entitled *Projects on the Road* or *Summer*, or even *Christmas*, have been designed to complement real life, amicably combining the memory of what has passed with readers' own daily life. In fact, the posts offer images and captions taken from the archives that match real life, for example, showing the Christmas windows of la Rinascente in December or hints of summer trips, wishing people happy holidays, and accompanying them discreetly.

---

[22] Sambonet S.p.A. was founded in 1856 by Giuseppe Sambonet, a graduate of the Academy of Fine Arts and a Master Goldsmith; in the 1930s, a plant was built capable of industrially producing solid silver cutlery flanked by galvanized silver; steel production also began. In 1947, the company began to produce stainless steel cutlery using its own technology; a few years later, in 1954, Roberto Sambonet, who had returned from Brazil and was working for the Rinascente company, began to collaborate with his father and uncle's company, exhibiting at the 11th, 13th and 14th Triennials.

Fig. 4 The post reads: "This August, a month of holidays and travel, we will take you with us to discover some projects kept in our archives and made in foreign countries or that show an international scope". The 1954 exhibition dedicated to Japan by la Rinascente, when Amneris Lis was the art director

4 Conclusion

The posts dedicated to Made in Italy on Facebook are clear indicators of CASVA's approach to communication, whose keys to interpretation are basically three: fidelity to rigorous scientific and methodological premises whose roots lie in the study of archival sources; an interweaving and balance between analogue and digital activities; and a desire to participate in the life of donors, collaborators and readers alike, bringing them into the picture even outside the workspaces. It is the CASVA itself which sums up these objectives in the invitation *Ami gli archivi? Vieni al CASVA!* (tat is, Do you love archives? Come to CASVA!) which, although aimed at civil servants, can be extended to everyone, in addition to those specifically interested in the Made in Italy archives.

Competing Interests The author has no conflicts of interest to declare that are relevant to the content of this chapter.

References

Carones M, Landsberger M (2006) 55/05. Cinquant'anni di professione. La provincia di Milano e i suoi architetti. Electa-Ordine degli Architetti, Milan
CASVA (2015) Milano Città Immaginata. Comune di Milano-CASVA, Milan
Dellapiana E (2022) Il design e l'invenzione del Made in Italy. Einaudi, Turin
Feraboli MT (2006) L'archivio dell'architetto Mario Terzaghi presso il CASVA. Comune di Milano-CASVA, Milan
Ferrario R (2024) Gli archivi come strumento di narrazione. In: Archivi del Contemporaneo. Atti del corso di Alta formazione. Nomos-MAGA, Busto Arsizio, pp 287–295
Fratelli M (2021) Un rinnovato inizio. In: Ciagà GL (ed) Gli archivi di architettura design e grafica in Lombardia. Mimesis, Milano-Udine, pp 6–7
Iannello M (2016) Roberto Sambonet artista e designer. Comune di Milano-CASVA, Milan
Irace F (1996) Milano moderna. Architettura e città nell'epoca della ricostruzione. Federico Motta Editore, Milan
Irace F (2017) The consumerism factory. In: Bandera A, Canella M (eds) R100 Rinascente. Stories of innovation. Skira, Milan, pp 159–167
Irace F (2021) Milano moderna: architettura, arte e città 1947–2021. 24 Ore Cultura, Milan
Pansera A, Abis M (1998) La fabbrica verticale. Il gruppo industriale Busnelli. Electa, Milan, p 26
Pica A (ed) (1951) Nona Triennale di Milano. SAME, Milan, pp 82–83
Quintavalle AC (1993) Design. Roberto Sambonet. Federico Motta Editore, Milano, p 13
Sacerdoti P (ed) (2020) Milano Città Immaginata: 10 progetti dagli archivi CASVA. Comune di Milano-CASVA, Milan
Schnapp J (2013) Animating the archive. In: Irace F, Ciagà L (eds) Animated archive. Electa, Milan, pp 63–71
Schnapp J, Mattei M (2015) Digital humanities. Egea, Milan

Valorization, Promotion and Communication of Made in Italy in the Archivio Progetti, Università Iuav di Venezia

Rosa Chiesa and Teresita Scalco

Abstract At the Università Iuav di Venezia belongs Archivio Progetti, a research and archival center, plays a pivotal role in the preservation and promotion of Italian design and craftsmanship culture, particularly through its extensive collections that include the works of Paolo De Poli, Giorgio Casali, and Luca Meda. These collections offer a comprehensive view of the evolution of Made in Italy, from the intersection of art, design, and photography to the broader cultural and economic implications of Italian creativity. De Poli's archive reflects his innovative contribution to enamel art; Casali's photographic archive documents key moments in Italian design, architecture, and urban transformations; Meda's collection highlights his synthesis of architecture and design, revealing his contributions to iconic furniture and industrial design.

The valorization of these archives involves a series of research projects, educational initiatives, exhibitions, and publications aimed at deepening the understanding of Italian design history, which include addition to the ApOnline, a digital strategy communication and the Petit Tour platform that offers an interactive and visual approach to digital heritage.

Keywords Outreach · Digital cultural heritage · Research tools · Exhibiting archives · Design archives

1 The Iuav Archivio Progetti

Archivio Progetti, the research and archival center of the Università Iuav di Venezia, is part of the Documentation and Library System; its collections relates to the design and architecture culture in its vast disciplinary articulations and scales: from

R. Chiesa (✉) · T. Scalco (✉)
Università Iuav di Venezia, Venice, Italy
e-mail: chiesa@iuav.it; tescalco@iuav.it

P. Cordera (ed.), *Unfolding Made in Italy (1948-1962)*, Research for Development, https://doi.org/10.1007/978-3-032-10118-1_5

Fig. 1 The historic headquarters of the Archivio Progetti in Venice, featuring Massimo Scolari's iconic "Wings". Università Iuav di Venezia/Umberto Ferro

landscape architecture, urban planning, until furniture and industrial design, graphic design, photography, visual arts and visual communication of the twentieth and twenty-first centuries.[1]

Archivio Progetti Iuav serves as a shared space for the scientific community, nationally and internationally, by making available primary sources (authorship drawings, grey literature, unpublished writings, models collections) to balance preservation, management and scholarly enhancement of documentary heritage, by creating research tools, by publishing its outputs and dissemination its heritage in exhibitions, both physical and online (Lupano and Tonicello 2018).

During its initial phase of activity, the Archivio Progetti was primarily focused on managing materials used and produced for architecture exhibitions. After several years at its original location in 1991, the Archivio Progetti moved to the former Venetian cotton mill, an industrial building restored by architect Gino Valle and mainly dedicated to educational activities. With a growing number of spaces, staff and funding, the Archivio Progetti has expanded its activities accordingly and in 2024 it has transferred its offices and collections to Magazzino 6 Ligabue, significantly increasing its spaces, which include in the basement the open model gallery, slide library, and compact storage area where photographic collections are kept, at the main floor the public services offices where the collections are displayed, and the second floor the digital laboratory with more drawings and materials (Fig. 1).

[1] On alternative approaches to designers' archives, see Feraboli's essay in this volume.

2 Beyond Conservation: Archive and Enhancement

The initiatives carried out by the Archivio Progetti are many and varied, covering a range of areas that reflect its primary mission as a place dedicated to conservation and research, but also to the dissemination of this research, in both analogue and digital form, through traditional tools such as publications, exhibitions and conferences, as well as innovative tools that embrace the philosophy of greater accessibility and knowledge of the collections preserved (Paalman et al. 2021).

A limited selection of three case studies, presented below and specifically related to the field of design and Made in Italy, shows how the Archivio Progetti has been both a driving force and an indispensable resource for the realisation of numerous valuable scientific activities.

The first archive acquired by Archivio Progetti in 2001 devoted to design and crafts, it has been the hybrid fund by Paolo De Poli[2] who was a renowned Italian artist, designer and craftsman best known for his pioneering work in enamel on metal, in fact his archive reflects the dual activities, on one side the typical personal-professional paper production, intersects with the documentation produced by the De Poli firm, thus reflections the organization of the archive of a small workshop-business.

Paolo De Poli generously documents his artistic-creative journey as well as his entrepreneurial path as a craftsman-designer firmly rooted in the patavine territory, both by education and profession. Trained as a painter before turning to enamel art in the 1930s. He developed innovative techniques that elevated the use of enamel in decorative arts, collaborating with leading architects and artists, such as Gio Ponti, Melchiorre Bega, Filippo De Pisis, and Marcello Mascherini.

At the same time, he was able to engage in dialogue at the national and international levels, distinguishing himself for his tireless industrious, intellectual vitality, and vigorous commitment to defending the role of the craftsman in dialogue with the rise of design culture in Italy in the 1960s and 1970s.

De Poli's works, which bridges art, craft and design, earned international acclaim and were displayed in major global and prestigious exhibitions in Brussels (1935), Paris (1937) and in several editions of the Venice Biennale, and the Milan Triennale—that acted as essential showcases for the Italian creativity and further solidified his reputation as a representative of Italian design, decorative arts and the Made in Italy.

[2] Paolo De Poli (Padua, 1905–1996) was born in Padua, Italy. He trained in drawing and metal embossing at the Pietro Selvatico art school in Padua and studied oil painting in Verona. Initially he worked as a portrait and landscape painter, exhibiting for the first time at the Venice Biennale in 1926.

Of considerable size, the archival collection is an essential repository of documents useful to scholars and professionals who want to explore not only the portrait of Paolo De Poli, but also the numerous access points that allow for expanding studies on previously unexplored histories of the arts (decorative, applied, sacred), craftsmanship, design (of objects and interiors), economics and technological innovations, blending traditional craftsmanship with modern sensibilities, that De Poli brought into the art of grand feu enameling, often inspired my ceramics, as well documents with the relationship with the ceramist Pietro Melandri.

De Poli transformed his *bottega,* or workshop, into a model of the Italian small *ditta*, or enterprise, successfully merging artistic endeavors and international entrepreneurial vision. In fact, his approach contributed to a narrative in which Italian design is seen as both creative and commercially viable, underscoring the economic value of cultural production.

In this frame, the dissemination of De Poli's creations was part of an international wide system of activities supported by the ICE (Istituto per il Commercio Estero or Institute for Foreign Trade)—such as curated exhibitions, trade fairs, and cultural promotional campaigns, where his works were externally displayed.

With this legacy, the acquisition of the De Poli fond expanded the disciplinary boundaries of Archivio Progetti collections, originally predominantly devoted to architects, engineers, and urban planners, highlighting the pervasiveness of interdisciplinary relationships. Starting in 2013, a season of archival management, description, and research actions led to the realization of an educational exhibition, a conference, and the publication of two books: the exhibition *De Poli. Craftsman and designer*, and an education project curated by Alberto Bassi, Valeria Cafà, and Teresita Scalco, within the course *Contemporary design critic* (prof. Alberto Bassi), part of the master's degree in Product and visual communication design (academic year 2014–2015) (Fig. 2); and the following books edited by Bassi and Maffioletti (2018), and Bassi and Cafà (2019).

More recently, in 2025, for the 120th anniversary of his birth, the Università Iuav di Venezia signed a partnership with the Municipality of Padua, Department of Culture, creating a narrative exhibition journey called *Paolo De Poli. Il Progetto dello smalto tra laboratorio e impresa* (or Paolo De Poli. The enamel project: between workshop and enterprise), curated by Alessandra Bosco, Lucilla Calogero, and Elisabetta Gastaldi, that connects the core collection of artworks at Museo degli Eremitani with the original archival documents from Archivio Progetti.

Among the recognised masters of Italian design imagery, Giorgio Casali[3] stands out. He was a photographer whose collection was the only one purchased in the

[3] Giorgio Casali (Lodi, 1913-Milan, 1995) was an influential Italian photographer known primarily for his work documenting postwar Italian architecture and design. Casali moved to Milan in 1928

Fig. 2 Exhibition setup of the education display *De Poli. Artigiano e designer*, curated by Alberto Bassi, Valeria Cafà, and Teresita Scalco, Archivio Progetti Iuav, Venice, May–July 2014. Università Iuav di Venezia/AP

1980s by the Iuav Archivio Progetti. Consisting of approximately 150,000 items, including photographs, negatives, and printed material, it documents in particular architecture, urban transformations, and design—from interiors to products—relating to the photographer's 30 years of work for *Domus* magazine. The rich collection, now organised, is a valuable resource for its ability to witness the relationships forged through professional activity in collaboration with the Pontian magazine, with designers and producers from 1951 onwards. A second interesting aspect of the collection concerns the presence of photographs belonging to Casali's personal photographic research, portraiture, and travel reportage-style work, sometimes in neo-realist style, which demonstrate how Casali's professional activity was not separate from the passion he loved to cultivate in his private life. A first exhibition entitled *Giorgio Casali. Photographer/Domus 1951–1983. Architecture, design, and art in Italy*, curated by Maggi and Zannier (2013), both professors at Iuav at the time, was held in 2013 at the Iuav headquarters in Venice and subsequently at the Centro

---

and started his apprentice at the Rambaldi photographic studio in Milan, where he worked for over 10 years, learning technical skills and developing his photographic ability. In 1938, Casali started his own photographic studio in partnership with Giovanni Muzzarelli before eventually opening a studio under his own name around 1950.

Internazionale di Fotografia Scavi Scaligeri in Verona, accompanied by a book of the same name published by Silvana Editoriale in 2013. In September 2025, a second exhibition on Giorgio Casali opened, entitled *Giorgio Casali fotografo: Paesaggi pubblici e privati* (or Giorgio Casali photographer: public and private landscapes), curated by Archivio Progetti—Giovanni Marras (scientific director), Rosa Chiesa, and Angelo Maggi, all Iuav professors.

The exhibition aimed to offer an updated critical interpretation of the photographer's most significant work areas—design, portraits, architecture, productive landscapes, and reconstruction—to convey the complexity of his artistic personality through visual narratives linked to his professional activity and his more private life. In particular, the term "public and private landscapes" was used to emphasise his mastery in interpreting the construction of the new urban and social society, expressed in the domestic, representative, productive, and human imagery interpreted through photography. A special section was dedicated to portraiture, which includes well-known figures from the world of art and design, without forgetting anonymous "human reportage".

A second significant collection examined here is that of Luca Meda, a designer who, after studying at the Hochschule für Gestaltung in Ulm and initially collaborating with architect Marco Zanuso, was able to combine architecture and design in his ongoing research and intense professional activity, bringing them together in a harmonious and constant conceptual and figurative relationship (Chiesa and Califano 2022).

The collection, deposited in 2013 at the Archivio Progetti of the Università Iuav di Venezia, consists of a vast quantity of drawings, sketches, photographs, prototypes and models of design and architecture from the architect-designer's studio, was investigated through a research grant dedicated to its initial organisation and a new cycle of studies reserved for a leading figure in Italian and Milanese architecture and design since the 1960s. The work led to the development of two research and dissemination activities: an exhibition, *Luca Meda. La felicità del progetto*, staged at the Triennale in Milan in 2013, curated by Nicola Braghieri, Rosa Chiesa, Serena Maffioletti (at the time scientific director of the Archivio Progetti), and Chiara and Sofia Meda, which mainly showcased "Meda's design" through study drawings and quick sketches, alongside models, moulds, and prototypes created for companies such as Radiomarelli, Molteni&C, Girmi, Dada, and Unifor (Figs. 3 and 4).

The poetic capacity of the designer's expressive style also emerged in the 2021 catalogue, *Luca Meda, Architecture, Design, Drawings* (Braghieri et al. 2021), richly illustrated and accompanied by thematic critical essays which, broadening the focus on the designer's areas of interest, investigated based on archival materials, offered a comprehensive overview of his work and the state of the art of research on his legacy.

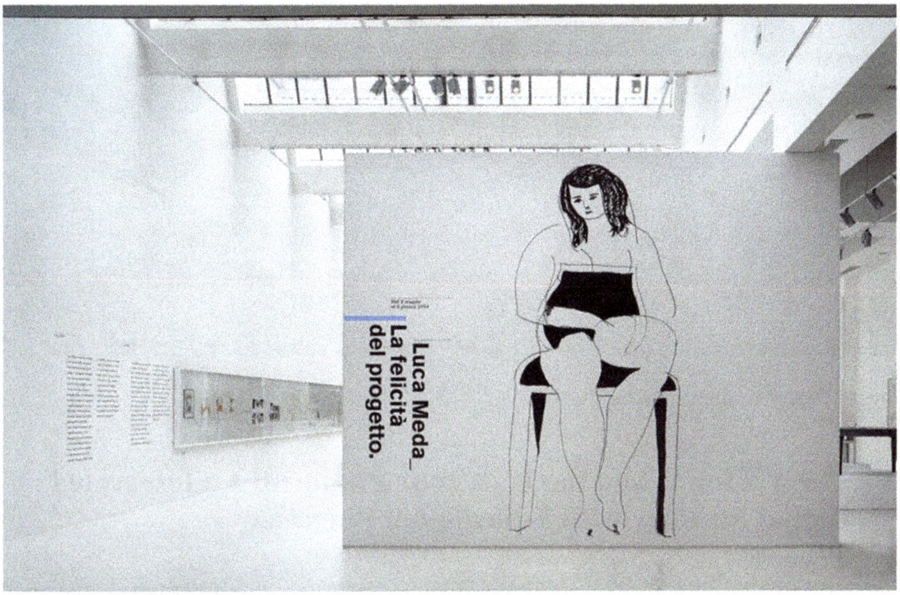

Figs. 3 and 4 Images from the exhibition *Luca Meda. La felicità del progetto*, curated by Nicola Braghieri, Rosa Chiesa, Serena Maffioletti and Sofia Meda, Triennale, Milan, May–June 2013. (Courtesy of Armando Bernacchi)

Fig. 5 Petit Tour digital platform

3 The ApOnline and the Petit Tour Platform: An Integrated Digital Strategy with Educational Purposes

A significant turning point in our digital strategy arose when UNESCO published the report *Toward Knowledge Societies for Peace and Sustainable Development* (UNESCO 2013). One of its key messages highlighted the crucial role of informal education and learning through digital content, especially via social media and internet-based platforms (Mansell and Tremblay 2013).[4] Responding to this call, since 2015, we have developed and launched the ApOnline integrated digital strategy. This initiative blends our evolving web presence with social media communication and a series of digital projects. The goal is to establish a dynamic, open system, centered on an editorial plan and a multimodal approach that fosters meaningful user interaction. We also draw upon the concept of 'extended visions' for collections, as articulated by Herbert Bayer (Perin 2020), to encourage a more participatory experience (Simon 2010).

Traditionally, access to cultural heritage has been limited to physical spaces, restricting public engagement and knowledge dissemination. In recent years, digital exhibitions have become a vital strategy for museums and cultural institutions to broaden their audiences and share their collections globally. These platforms offer a range of advances: they enable access by people who cannot visit in person, safeguard fragile documents, and extend the longevity and impact of public interaction with cultural materials.

To address these new scenarios, during the COVID-19 pandemic in 2020, we created Petit Tour, a platform of digital exhibitions, designed to maximize the visibility and accessibility of our collections (Fig. 5).

The platform employs the graphic approach of Instagram, supporting a highly visual, intuitive, and mobile-friendly user experience. Its adaptable design

[4] As a further example, see the essay by Ceconello and Spallazzo in this volume.

Fig. 6 Petit Tour edited by Anna Giambellini and Cecilia Marcolini created during the master's course *Metodologie della ricerca: fonti, archivi e reti* by Rosa Chiesa in 2022–2023

accommodates a range of curatorial and communication needs, ensuring that both thematic, monographic, and chronological displays can be created and made globally accessible.

With the launch of Petit Tour and our expanded ApOnline digital strategy, we have embraced the full potential of the digital sphere to share and safeguard our cultural heritage. This digital innovation fosters new avenues for learning, research, and creative reuse, reflecting an inclusive vision of academic practice and heritage preservation (Scalco 2025).

Our ongoing commitment to digital accessibility is informed by the understanding that open archives and digital exhibitions are active spaces for community engagement and educational activities, such as the involvement of the students from the *Metodologia della ricerca, archivi e reti* course of Rosa Chiesa. The course offered in the 2022–2023 academic year aimed to teach a historical method, consolidate the importance of historical sources for research, and enhance the written and visual communication of the research itself. The exercise proposed in collaboration with Archivio Progetti, carried out in small groups of diverse backgrounds (design and architecture curricula), aimed to introduce students to archival work and the physicality of sources, while requiring a digital artefact as the final output: a Petit Tour (Fig. 6).

Through these efforts, we continue to expand the impact and relevance of cultural heritage in the digital age, ensuring that our collections remain accessible, meaningful, and inspiring for future generations.

4 Conclusion

The experience of the Archivio Progetti can be seen as emblematic of the transformations that contemporary archives have undergone and continue to undergo; moving beyond the idea of a repository of heterogeneous sources at the service of design research, the digital archive (not simply a copy of the analogue archive) becomes an opportunity to share—even with non-specialist audiences—understandable cultural content, as demonstrated by the Petit Tour digital project (Armitage and Guldi 2016).

The enhancement of sources as objects for designing new research paths and new relationships, in particular, cannot fail to combine traditional methods—publications, conferences, and exhibitions—with new tools that derive from the potential of digital technologies (Marsh 2023). Tools evolve in a generative perspective for research, but also at the service of enjoyment (Manovich 2017).

The ease of use of the resources stored in archives, and the resulting ease of learning, embraces the logic of social inclusion and increased accessibility, also in quantitative terms.

Thinking of the archive as a "relational asset" capable of generating new narratives and suggesting possible design actions that combine the digital dimension with the advanced and innovative profile of design, as implemented in the aforementioned educational experiment, means relying on what Digital Humanities expert Jeffrey Schnapp refers to as "knowledge design" (Chiesa and Filippini 2022).

The constant evolution of the nature of sources, which are now sometimes digitally native, ultimately requires continuous reflection on the most effective ways of disseminating and understanding the Archive. In the specific case of the Archivio Progetti, the desire to embody the idea of a "living archive", opening up and increasingly encouraging research into its collections, is more relevant than ever. Since 2015, the claim "living archive" has become a hashtag on Archivio Progetti's social media and succinctly summarises not only its mission but also the concrete creative actions implemented in recent years.

Competing Interests The authors have no conflicts of interest to declare that are relevant to the content of this chapter.

References

Armitage D, Guldi J (2016) Manifesto per la storia: il ruolo del passato nel mondo d'oggi. Donzelli, Rome

Bassi A, Cafà V (2019) Paolo De Poli e Gio Ponti. L'artigiano-designer e l'architetto. Universalia, Pordenone

Bassi A, Maffioletti S (2018) Paolo De Poli. Artigiano, imprenditore e designer. Il Poligrafo, Padua

Braghieri N, Carboni S, Maffioletti S (2021) Luca Meda, architecture, design, drawings. Silvana Editoriale, Milan

Chiesa R, Califano P (2022) Oggetti narrati e immaginati: Luca Meda e il disegno (Narrated and imagined objects. Luca meda and drawing), diségno, vol 11, pp 67–78. https://doi.org/10.26375/disegno.11.2022.9

Chiesa R, Filippini A (2022) Suggerire nuove storie dai documenti: l'Archivio Vico Magistretti. Percorsi di ricerca e networking tra archivi del design. Officina, Conegliano

Lupano M, Tonicello A (2018) Il movimento per gli archivi di architettura. Università Iuav di Venezia, Venice

Maggi A, Zannier I (2013) Giorgio Casali. Photographer/Domus 1951-1983. In: Architecture, design, and art in Italy. Silvana Editoriale, Milan

Manovich L (2017) Cultural data. Possibilities and limitations of the digital data universe. http://manovich.net/index.php/projects/cultural-data. Accessed 10 Jun 2025

Mansell R, Tremblay G (2013) Renewing the knowledge societies vision for peace and sustainable development. UNESCO, Paris. https://unesdoc.unesco.org/ark:/48223/pf0000224531?posInSet=1&queryId=204ed538-9185-458d-b9ec-9b77e80ed6fd. Accessed 20 Aug 2025

Marsh E (2023) Creating digital exhibitions for cultural institutions. A guide. Routledge, London

Paalman F, Fossati G, Masson E (2021) Introduction: activating the archive. The moving image. J Assoc Moving Image Archivists 21(1–2):1–25. https://doi.org/10.5749/movingimage.21.1-2.0001

Perin R (2020). Exhibition design and the relationship with the spectator: historical notes with El Lissitzky and Herbert Bayer. In: ICDHS12: lessons to learn? Past design experiences and contemporary design practices. Proceedings of the 12th international conference on design history and design studies, Zagreb (Croatia), October 2020. UP12M Books, Zagreb, pp 29–38. https://doi.org/10.29327/174570.12-1

Scalco T (2025) Educate about beauty: petit tour, online exhibitions as a means of designing a more democratic cultural heritage. In: Renaville F et al (eds) Opening up our heritage. Challenges and opportunities in digitising and promoting cultural and research collections. Presses Universitaires de Liège, Liège. https://doi.org/10.25518/978-2-87019-330-3.17

Simon N (2010) The participatory museum. Museum 2.0, Santa Cruz. https://participatorymuseum.org/. Accessed 20 Aug 2025

UNESCO (2013) Towards knowledge societies for peace and sustainable development. UNESCO, Paris. (published 2014). Available via UNESDOC, Digital library. https://unesdoc.unesco.org/ark:/48223/pf0000227146. Accessed 20 Aug 2025

The Tale of Made in Italy and the Massimo and Sonia Cirulli Foundation: The Relationship Between Collections and Exhibition Choices

Sandra Costa and Irene Di Pietro

Abstract The Cirulli Foundation was established in Bologna in February 2015 as the result of an important 30-year acquisitions activity aimed at enhancing Italian visual culture from the early twentieth century to the years of the economic boom. The Foundation's collection is often regarded as the most important privately held collection devoted to post-World War II graphic art and design in Italy. The essay aims to explore the museographic and museological aspects of the collection: the building (originally created by Achille and Pier Giacomo Castiglioni for Dino Gavina) and its philological restoration, the exhibition of designed objects created as an "animated archive" of thematic explorations (the "capsule collections"), developed by US designer Jeffrey Schnapp, the collection and the temporary exhibitions (following a complementary narrative methodology, combining cues, storytelling, and scholarly research) and its enhancement and mediation.

Keywords Archives · Collectionism · Design · Museography · Twentieth century

1 Materials and Curatorial Choices in Setting up the Collection

The Massimo and Sonia Cirulli Foundation was established in Bologna in February 2015 as the result of an important 30-year acquisitions activity aimed at enhancing Italian visual culture from the early twentieth century to the years of the economic boom. Today, the Foundation's collection—which is self-financed and guided by a scientific committee composed of Pierpaolo Antonello, Silvia Evangelisti, Nicola Lucchi, Ara H. Merjian, Marco Sammicheli, and Jeffrey T. Schnapp—is often regarded as the most important privately held collection devoted to post-World War

S. Costa (✉) · I. Di Pietro
Alma Mater Studiorum, Università di Bologna, Bologna, Italy
e-mail: sandra.costa@unibo.it; irene.dipietro2@unibo.it

P. Cordera (ed.), *Unfolding Made in Italy (1948-1962)*, Research for Development, https://doi.org/10.1007/978-3-032-10118-1_6

II graphic art and design in Italy. The New York origins of this endeavor, which began in 1984 under the name Massimo and Sonia Cirulli Archive, are evident in a series of conceptual approaches that are as revealing of curatorial practices as they are of exhibition design choices. These are both characterized by dynamic planning and by the idea of the "open work." Moving away from traditional museographic formats, the Foundation embraces a narrative dimension, aiming to offer the public a reorganization of materials through a network of relationships, chosen to support a multidisciplinary storytelling of the roots of Made in Italy with an international outlook. What is proposed is a narrative that is attentive to the complexity of the visual history of the twentieth century and that seeks to highlight, through the transversality of knowledge and practices, the richness of Italian creativity. The display adopted leverages the unique characteristics of a collection that, *ab origine*, sought to avoid hierarchies between high and low art, between decorative arts and industrial design, between auteur graphics and visual communication. For example, design drawings from the 1930s to the 1970s trace the studies of artists, architects, and designers of the period. In addition, a rich collection of posters features the work of leading Italian graphic designers and illustrators, often linked to the visual communication of companies such as Alemagna, Barilla, Campari, Fiat, Olivetti, and Pirelli—firms that have played a central role in the history of Italian industry and Made in Italy, and whose advertisements are now recognized as possessing their own artistic value.

2 Spaces and Collections

The Foundation's headquarters were inaugurated in 2018 in San Lazzaro, near Bologna, alongside the opening of its first exhibition, *Universo Futurista*, curated by Jeffrey Schnapp and Silvia Evangelisti. The site chosen by Massimo and Sonia Cirulli for the Foundation is an iconic space tied to Italian architecture and design of the 1960s. It was originally created by designers Achille and Pier Giacomo Castiglioni for Dino Gavina, who had transformed his furniture factory into a meeting place for designers and artists such as Carlo and Tobia Scarpa, Marcel Breuer, Man Ray, Lucio Fontana, and Sebastian Matta. This was, in fact, the first realization of a "laboratory of ideas"—a concept that still characterizes the site today. For example, in one of the warehouses not yet restored, visitors can discover an underground space that continues to be used as a venue for exhibition events.

The Castiglioni project drew inspiration from the rural architecture of Emilian barns. In issue 385 of *Domus* (Edificio per mostra 1961), it was described as "a space invented with freedom and rigor." When the building was acquired by the Foundation, it was in a very poor state of repair. Its restoration was carried out with an almost philological approach by Elisabetta Terragni (from Studio Terragni Architetti, based in Como and New York) and Daniele Ledda (from XyComm, Milan), who stated their intention was to "follow the Castiglioni on tiptoe," aiming to fully respect the original space.

Elisabetta Terragni preserved the existing features, such as the terracotta floors, the cast-iron radiators, and the window frames, intervening only where necessary to bring the rooms up to standard and using methods in line with the original design.

In this case, the architectural space, both before and after restoration, is itself a carrier of meaning in relation to the idea of "staging" Made in Italy. Today, the relationship between container, content, and the display of the collections offers a valuable perspective on the distinctiveness of the exhibition choices reflected in the layout.

From a curatorial point of view, the guiding concept behind the Foundation's display and cultural strategies has been that of the "animated archive", developed by US designer Jeffrey Schnapp. Since the late twentieth century, the relationship between archives and museums has become increasingly close, as both institutions share a dual nature, functioning as spaces of both preservation and production of memory (Zuliani 2014).

A key moment in the theoretical reflection on the archive as an institution, and on the evolving relationship between archives and exhibitions, occurred in 1995 with Jacques Derrida's *Mal d'Archives*, whose insights have continued to influence discourse into the twenty-first century.[1]

While persistent stereotypes still depict archives as closed, orderly, and silent "time boxes," in reality, their purposes and methods of narrating contemporaneity are increasingly polymorphous, dynamic, and permeable.

Written sources—as well as graphic, photographic, and now digital ones—make up a body of work that transforms the archive into a place of interdisciplinary correspondences, open to diachronic and synchronic inquiry. Georges Didi-Huberman was probably one of the intellectuals who contributed most effectively to the spread of a broader conception of the archive, conceived as both a creative and museological procedure, involving the production and management of knowledge and memory. Didi-Huberman drew on the myth of Atlas—and notably on the Warburgian model—to evoke a "visual form of knowledge" capable of composing, decomposing, and recomposing memory, a necessary process for forging a relationship with History and histories (Didi-Huberman 2010; Trione 2011; Baldacci 2013).

To follow Jeffrey Schnapp's methodological proposal, it is useful to move beyond the traditional distinctions between collection and archive found in manuals, and to recall, once again, Aby Warburg (1866–1929) and his famous *Mnemosyne Atlas*.

Since the late twentieth century, critical reflection on archives has been closely linked to experiments related to contemporary creativity, engaging artists who incorporate the theme of the archive into their poetics as a privileged tool for exploring the interpretation of reality (Enwezor 2008; Foster 2004; Osthof 2009).

According to Schnapp, the animated archive developed for the Cirulli Foundation is conceived as a space of experience, where visitors encounter the collection's materials through a kind of conceptual continuum: between the archive as a site of

[1] For another perspective on the use of archives, see the essays by Feraboli and by Chiesa and Scalco in this volume.

preservation and memory, and the exhibition pathway as a space of sharing and activation. What Schnapp calls "a no man's land" becomes the terrain for a multifaceted and constantly evolving narrative—one that centers on a material culture that is heterogeneous in materials, techniques, and styles, but also in the symbolic and cultural values it conveys.

The collection presented in the Foundation's spaces aims to serve as a cross-section of twentieth-century Italian culture with its ambiguities and contradictions. This form of memory is conceptually attentive to the research conducted by historians of the Annales school, which led to the valorization of micro-histories and otherness, thereby promoting a conception of history defined by intentionally broader boundaries.

From a theoretical perspective, the animated archive presents itself as a kaleidoscopic mosaic of materials that easily traverse the boundary between high art and mass culture, offering representations that include figures and objects often considered "minor." What takes shape through this concept is a vision of twentieth-century history and culture that is more global and more attuned to an expansive notion of heritage. In short, it is also—and perhaps above all—a deliberate rethinking of traditional disciplinary and collecting divisions through a kind of *métissage* between different forms of visual culture.

This idea of an animated archive also envisions a "dynamic" type of exhibition in which selected objects may be periodically rotated, offering different nuances to the narrative of an era. Just as in the "open work," following Umberto Eco's well-known definition, art is transformed from object to process, so too in the exhibition does the staging of an animated archive shift the focus from product to process, and from solutions to questions (Costa and Ruozzi 2020).

Even the exhibition design reflects the world of the archive: for example, through the use of grids that are the same in the exhibition halls and in the archive areas. The archive, like the museum, thus confronts its most significant challenge, which is not related to the consultation or preservation of data, but to the ethical question of the meaning we assign to them. How can we forget, for instance, that access to the archive and the interpretation of its contents becomes, according to intellectuals such as Jacques Derrida, a possible measure of democracy? (Derrida 1996; Foucault 1969).

3 The Museographic Dimension

The transition of the Castiglioni-designed building from Gavina's designer-furniture showroom to the Cirulli Foundation's headquarters led not only to a change of its intended use but also to a form of "musealization" of the space.

During the philological restoration, the architect made several bold interventions that doubled as museographic strategies. For example, the mandatory addition of railings and public-access stairs was combined with an expansion of the exhibition area. The "showcasing"—almost as if it were a work of art—of Castiglioni's staircase, along with the construction of entrance display cases using the same black steel profiles adopted by Achille and Pier Giacomo Castiglioni and designed to avoid creating a visual void in relation to the floor below, marked the establishment of a "stable" museographic dimension within the space. Meanwhile, the large glass areas serve as ephemeral exhibition surfaces, regularly featuring temporary window-graphic installations.

From the point of view of spatial articulation, perhaps the most significant aspect is the decision to use the same black grids in the exhibition rooms as those adopted in the archive reserves. In this way, the exhibition discourse and the theoretical discourse are jointly developed through a preference for a dynamic museographic choice that seems to enable the transformation of the temporary into the permanent and vice versa and allows for the easy replacement of parts of the display by modifying—or updating with new materials and new acquisitions—the narrative proposed by the Foundation.

The staging and the use of black archival grids thus become the most immediate visual expression of a different hierarchy between Art and applied arts, and through this intentionally smart and flexible setup, a story of design, conceived in its broadest sense, comes to life. In the various sections, each characterized by a distinct color, the exhibition unfolds through narrative poles developed across multiple domains: architecture, industry, fashion and design, speed and flight—so beloved of Futurism—and propaganda. Each section bears a simplified alphanumeric code as its title, recalling the system used in the archive.

In the context dedicated to Made in Italy and the post-World War II years, the example offered by section AA02TM—that is, Animated Archive, Section 02, Fabrics of Modernity—may be of interest. The narrative proposed is that of the transition from the tradition of artisanal manufacturing in the early twentieth century to industrial production and textile design in the 1950s, by artists and designers such as Lucio Fontana, Piero Dorazio, Enzo Mari, and Fausto Melotti. Drawings and sketches made for the 10th and 11th Milan Triennale (1954 and 1957) help the audience understand the role that textiles and textile companies played in the spread of an Italian style throughout the world.

Also defining the character of the 1960s is a focus on creativity related to plastic objects and design. This ductile and inexpensive material entered everyday life by making accessible many items that would have been too costly if produced using traditional techniques and materials, while also supporting new forms of consumption—such as the iconic Radio Cubo designed by Marco Zanuso for Brionvega in 1963. With the transition to the 1970s and 1980s, space is devoted to the creative universe and pop style of fashion designer Elio Fiorucci.

In its displays and its policy of enhancing the collections, the Foundation has moved away from the primacy of the "masterpiece" to demonstrate how, after World War II, a form of communication capable of using the voice of objects developed to propose new consumption, different ways of life, and modern sociability. Attention is also given to the field of corporate advertising, which, as in the case of Pirelli, became a meeting place for some of the best-known Italian and international designers.

In the Cirulli Foundation's exhibits, collections coexist to tell the story of twentieth-century Italian society in complementary ways, and design is placed within a broader context of meaning. The aim is not so much to present and give voice to a single object, even if iconic, but rather to show, through a collective vision, how visual strategies and everyday items—born from the creativity of Italian artists and designers—contributed to changing tastes and ways of living, thanks to the international affirmation of Made in Italy.

From the Foundation's front door, Walter Gropius's quote from the preface of the Italian edition of *Scope of Total Architecture* (Gropius 1959) sets the tone for the spaces and offers the public the first key to look at, but more importantly to rethink, the collection and its exhibits: "Perhaps Italy is destined to clarify what data of modern life we must base ourselves on in order to recover the lost sense of beauty and promote in the industrialized era a new cultural life" (Fig. 1).

Fig. 1 The Massimo and Sonia Cirulli Foundation's front door with Walter Gropius's quote from the preface of the Italian edition of *Scope of Total Architecture*

4 Enhancement and Mediation

Massimo and Sonia Cirulli's choice to place their collection in the Gavina showroom is deeply connected to the idea of opening this rich heritage to the public so that its aesthetic and cultural dimensions can be experienced, while also allowing the Foundation's ongoing research activities to continue. The intent to promote accessibility and visibility of the works is linked both to selecting a venue already defined by its exhibition vocation and to fostering dialogue with visitors through ever-changing exhibitions (Serri 2018; Cirulli 2024). In both large and small exhibitions, the Foundation offers thematic explorations that invite viewers to focus on fragments of twentieth-century industrial art culture.

The commitment to publicizing Made in Italy as the basis of modern culture has led the Cirulli Archive to collaborate with prestigious international institutions by loaning works for temporary exhibitions: the Guggenheim, for which it contributed to *Italian Futurism, 1909–1944* held in New York and Bilbao in 2014; the De Young Museum in San Francisco, with which it collaborated on *Jewel City—Art of the Panama-Pacific International Exposition* in 2015, including Giacomo Balla's *Disgregazione x velocità, Penetrazioni dinamiche d'automobile*—one of the very few works exhibited in the Futurist gallery of the 1915 Pan-American International Exposition. For the Musée d'Orsay, the Foundation participated in the exhibition *Dolce Vita? Art décoratif italien 1900–1940, du Liberty au design industriel* in 2015. Collaborations with the Ferragamo Foundation include *Tra Arte e Moda* in 2017, and other exhibitions such as *1927 Il Ritorno in Italia*; *L'Italia a Hollywood* (2018), and *Donne in Equilibrio (1955–1965)* in 2023 at the Ferragamo Museum. Finally, it worked with Fondazione Prada on the 2018 exhibition *Post Zang Tumb Tuuum. Art Life Politics: Italy 1918–1943*, curated by Germano Celant.[2] The Cirulli Foundation Archive is also part of the new exhibition routes at the Museum of Italian Design at Triennale Milano.

Above all, however, it is in San Lazzaro that visitors can enter a dynamic and ever-changing space where they are accompanied through the suggestions arising from the works displayed within time capsules designed by Jeffrey Schnapp that are renewed cyclically. These capsules function as self-contained, replaceable, and renewable micro-exhibitions tailored to the evolving needs of displaying the entire collection and its narrative.

The display's organization is minimal; rather than captions or tags (which are completely absent), it is the color palettes associated with each exhibition section that suggest connections between objects (Fig. 2).

---

[2] For issues concerning design items on display, see the essay by Carella, Zurlo, and Martino in this volume.

Fig. 2 Massimo and Sonia Cirulli Foundation. Capsule AA14MW, dedicated to the *Monotypes of Venturino Venturi*

Due to the diversity and variety of works, the Foundation encourages "a kaleidoscopic approach" (Schnapp 2019).

Regarding the challenge of enhancing the value of its heritage, the Foundation aims to move beyond the initial curiosity sparked by the Castiglioni building's architectural form, inviting visitors past the threshold and into spaces where the collection can be fully experienced—offering a cultural journey that draws them closer to the themes presented and helps them grasp the interdisciplinary connections suggested (Fig. 3).

Fig. 3 Massimo and Sonia Cirulli Foundation. Second floor of Achille Castiglioni's building

5 Temporary and Traveling Exhibitions

The exhibitions proposed by the Foundation follow a complementary narrative methodology, combining cues, storytelling, and scholarly research. Italian culture up to the period of the economic boom is explored with an analytical lens through capsule collections and traveling exhibitions. However, the boundaries between these narrative formats are not rigid. The first temporary exhibition, *Universo Futurista*, for instance, was conceived as an ideal complement to *Universo Futurista. 11 Futurist Satellites*, held at the Archiginnasio Municipal Library in Bologna. The same thematic core also led to *Futurist Universe*, organized in collaboration with the Tsinghua University Art Museum in Beijing in late 2022. Some sections currently on view within the Foundation have evolved from earlier capsule exhibitions over the years: for example, the small thematic displays *Fiorucci Pop Revolution* (Fig. 4), *Federico Fellini Behind the Scenes*, and *The Women of Kabul. In the Gaze of Photographer Pino Settanni* (16 March to 31 December 2022). Another case is the exhibition *Cut&Paste. Photocollage between Dada and*

Fig. 4 Massimo and Sonia Cirulli Foundation. Capsule dedicated to *Fiorucci: pop revolution*

Futurism, which opened in March 2023 as part of the 46th edition of Arte Fiera in Bologna.

In addition, temporary exhibitions have been installed within the Cirulli Factory, such as those presented in the rooms of the former Gavina warehouses, which have not yet been restored or repurposed and are open to the public only for limited periods.

The Foundation's Scientific Committee may contribute to the development of these temporary exhibitions. Research is supported by a digital database, allowing curators to consult the entire collection and begin their study of individual objects. Temporary traveling exhibitions, which can be loaned as a single corpus, are also critically developed by the Foundation, and essays are provided alongside them. These materials can be used by the hosting institution to curate the exhibition in its own spaces, potentially presenting different outcomes.

The Birth of Made in Italy presents a vision of 1950s Italy through its modern, industrial development. The exhibition includes over 200 design objects by more than 30 artists, together with advertising graphics and photography. The so-called *economic miracle* is depicted through major transformations: changes in lifestyle, innovative communication campaigns by prominent figures, and the creation of iconic branded products.

This exhibition formed the core of *Italy of the Miracle. Art, Design and Advertising 1948–1968*, curated by Elisa Fulco, Sonia Pellegrini, and Giuseppe Virelli (Fulco et al. 2014), which was held in the Church of Sant'Agostino and the Piersanti Museum in Matelica. The show retraced the development of Italian industry through the entrepreneurial story of Enrico Mattei, himself a native of Matelica (Orazi 2014).

In this first "exportation," Cirulli's works were conceived in dialogue with Baroque settings. In contrast, for the subsequent 2017 project *A forma das ideias. Design italiano do pós-guerra* (Instituto Tomie Ohtake and Fondazione Massimo e Sonia Cirulli 2017), held at the Instituto Tomie Ohtake in São Paulo, Brazil, the objects were displayed within a contemporary space specifically designed for exhibitions of art, architecture, and design.

The show *Lo Stile Italiano fra arte e design* (Cirulli and Cirulli 2011) focused on recounting the story of major companies and the construction of the idea of Made in Italy through communication campaigns and iconic objects, spanning from the birth of Modernity to the postwar economic boom and intertwining art, fashion, architecture, publishing, and design.

The exhibition was sponsored by the Italian Embassy as part of the celebrations for the 150th anniversary of the Unification of Italy and was first staged in 2011 at the Grimaldi Forum in the Principality of Monaco. It was later presented in 2013 at the Reggia di Colorno under the title *Lo stile Italiano. Art and Society 1900–1940.* Once again, the objects were installed in markedly different exhibition environments: from the glass-and-steel architecture of the Monegasque Culture and Convention Center to the Baroque interiors of the Ducal Palace of Parma, where the use of black plasterboard panels interrupted a potential dialogue between the exhibited objects and the original spatial context.

6 The Approach to Mediation

In the Cirulli Foundation's vision, the arrangement of the collection, as already mentioned, is also a functional tool for understanding the preserved works. However, this goal is not yet fully realized, mainly due to limited public access: visits are possible only during monthly openings. On these occasions, visitors can follow a guided tour with focused insights, based on the remodeling of the time capsule spaces. The Foundation considers in-person mediation the most effective way to engage with its audience.

The difficulty in increasing public openings stems primarily from the need to dedicate resources to the Foundation's multiple activities—such as conservation, archiving, international collaborations, and the buying and selling of works—which also support its self-financing efforts, as the Foundation does not benefit from public funding.

Mediation is therefore tied to the idea of opening the Foundation as a necessary step toward educating younger generations and foreigners who may be less familiar with Italian art. This perspective highlights the strategic importance of the Foundation's educational programs and offerings for schools. Following an edutainment approach, children are encouraged to critically interpret reality through art. Starting from the observation of the collection's works and their production contexts, a series of workshop itineraries, called *Orbits*, have been developed to explore the collection's cross-cutting themes: advertising graphics, flight, and notable designers such as Fiorucci and Munari. Each project includes a visit to the collections followed by a hands-on workshop. Inspired by the objects, participants create personal works such as graphic prototypes, advertising posters, watercolor postcards, and clothing collections using various patterns.

Schools in San Lazzaro di Savena are encouraged to take part in these programs, benefiting from municipal funding that allows them to participate free of charge. This helps build a valuable connection with the local community and younger generations.

7 What Storytelling Perspectives?

The heterogeneity of the materials, the very idea of an animated archive, and the modular composition of the installation all highlight the importance of building a dialogue between the works, while also offering valuable cues for investigating Italian material culture. The time capsules could potentially expand the range of their contents.

Indeed, Jeffrey Schnapp emphasizes that objects within an archive are preserved because they offer an integral view of historical memory and that archives enable us to consider a work and explore it in its diversity (Schnapp 2019). Within the collection, a digital strategy could be proposed and developed to amplify the meanings of the objects. At a time in history when the taxonomy of archives, libraries, and museums is undergoing profound change, as the curator theorizes, it becomes necessary to imagine a reflection that integrates these three institutions into the process of cultural production. The Foundation could embrace digital tools as a field of experimentation by showing the processes that generated the preserved works, allowing users to access and compare further productions by artists and designers, and enhancing understanding through multimedia resources. The possibility of exploring and creating references to the culture in which the works are embedded could be investigated through the use of the internet, especially considering that the Cirulli Foundation holds categories of objects that may evoke affection and a sense of closeness in some segments of the public.

Possible collaborations and dialogues with other institutions and archives may represent a viable path for openness, both physically and culturally, effectively enriching the relationship between users and the collection.

Competing Interests The authors have no conflicts of interest to declare that are relevant to the content of this chapter.

Authorship Attribution This essay is the result of collaborative work. Sections 1–3 were written by Sandra Costa, and Sects. 4–7 by Irene Di Pietro. The translation is by Alessandro Paolo Lena.

References

Baldacci C (2013) Mal d'archivio. La Repubblica
Cirulli M (2024) Cirulli Foundation. The Pirate Cove. Archivio 10: 80–81
Cirulli M, Cirulli S (2011) Lo stile italiano. Arte e design. Silvana Editoriale, Cinisello Balsamo
Costa S, Ruozzi G (2020) Prefazione. Teca 1:9–14. https://doi.org/10.6092/issn.2240-3604/11675
Derrida J (1996) Mal d'archivio. Un'impressione freudiana. Filema, Napoli
Didi-Huberman G (2010) Atlas ¿Como llevar el mundo a cuestas? Paper presented at the Museo Nacional Centro de Arte Reina Sofía, Madrid
Edificio per mostra (1961) Edificio per mostra di mobili sulla via Emilia, a Bologna. Domus 385:5–10
Enwezor O (2008) Archive fever. Uses of the document in contemporary art. Steidl/ICP, Göttingen
Foster H (2004) An archival impulse. October 110:3–22
Foucault M (1969) L'archéologie du savoir. Gallimard, Paris
Fulco E et al (2014) Italia del miracolo. Arte, design e pubblicità 1948–1968. Quattroemme, Perugia
Gropius W (1959) Architettura Integrata. A. Mondadori, Milano
Instituto Tomie Ohtake, Fondazione Massimo e Sonia Cirulli (2017) A forma das ideias: design italiano do pós-guerra: exhibition organized by Instituto Tomie Ohtake in collaboration with Fondazione Massimo e Sonia Cirulli. Instituto Tomie Ohtake, São Paulo
Orazi M (2014) Arte, design e pubblicità "L'Italia del miracolo" in mostra a Matelica. https://www.cronachemaceratesi.it/2014/07/29/arte-design-e-pubblicita-litalia-del-miracolo-in-mostra-a-matelica/556646/. Accessed 17 Jul 2025
Osthof S (2009) Performing the archive. The transformation of the archive in contemporary art from repository of documents to art medium. Atropos, New York
Schnapp J (2019) L'archivio animato, Conversation with Jeffrey Schnapp https://youtu.be/cV-c74Ti960?si=AaAYBB6zKib5ub_d. Accessed 17 Jul 2025

Serri C (2018) Fondazione Massimo e Sonia Airulli: Arte italiana dal primo ‘900 al boom economico. vietate le mostre “blockbuster”, Espoarte 18 September 2018. https://www.espoarte.net/arte/fondazione-massimo-e-sonia-cirulli-arte-italiana-dal-primo-900-al-boom-economico-vietate-le-mostre-blockbuster/. Accessed 17 Jul 2025

Trione V (2011) Mappe per la memoria. Corriere della Sera Cultura

Zuliani S (2014) “Là dove le cose cominciano” Archivi e musei del tempo presente. Ricerche di S/Confine 3: 81–89

Made in Italy on Display Inside and Outside Museums

Identitalia. The Iconic Italian Brands: Notes on Exhibition Challenges

Gianluca Carella, Francesco Zurlo, and Carlo Martino

Abstract This essay explores the curatorial strategy, research methodology, and cultural implications of *Identitalia. The Iconic Italian Brands*, an exhibition that showcases over 100 iconic Italian companies as expressions of national identity. Through the lens of design and public history, the chapter examines how historical sources—archival materials, product evolutions, corporate narratives—can serve as tools for constructing a collective memory rooted in industrial and cultural heritage. Emphasis is placed on the importance of material culture in narrating a nation's identity and on the role of exhibitions as mediating devices between past, present, and future. The research process combined digital ethnography and direct engagement with companies, revealing three typologies of archival culture—Ideal, Ambiguous, and Dormant—that shaped the curatorial approach. The chapter also discusses the main challenges encountered, such as disparities in corporate archival practices, restricted digital access, and the tension between historical transparency and brand self-perception. In response, the team adopted a flexible and inclusive strategy that turned constraints into narrative opportunities. The *Identitalia* exhibition becomes a case study for reflecting on how design exhibitions can foster recognition, promote historical awareness, and renew the discourse on national identity in the contemporary age.

Keywords Italian brands · Design and industrial heritage · Curatorial and exhibition strategies · Strategic design for corporate memory · Value expression

---

G. Carella (✉)
Politecnico di Milano, Milan, Italy

MIT Morningside Academy for Design-Massachusetts Institute of Technology, Cambridge, MA, USA
e-mail: gianluca.carella@polimi.it

F. Zurlo
Politecnico di Milano, Milan, Italy

C. Martino
Sapienza Università di Roma, Rome, Italy

P. Cordera (ed.), *Unfolding Made in Italy (1948-1962)*, Research for Development, https://doi.org/10.1007/978-3-032-10118-1_7

1 The Role of Historical Sources in Representing the National Identity

Exhibiting the essence of a nation requires more than a collection of artifacts; it calls for a coherent narrative rooted in time and place. In the case of Italy, this narrative is tightly interwoven with its iconic brands. These brands do not merely represent commercial success; they serve as cultural artifacts, traces of national evolution, and signifiers of collective memory. Historical sources, such as archival documents, advertisements, packaging, company records, and product design evolutions, become indispensable in reconstructing these trajectories.

Scholars in museum studies and design history have long underlined the importance of primary sources in articulating narratives of national identity (Pearce 1990; Kavanagh 2005). Exhibitions, particularly those addressing design and industrial culture, act as mediators between the past and present, offering anchored and anticipatory interpretations. As Susan Pearce (1990) suggests, museums serve as spaces where knowledge is created through the curation of objects, and these objects derive their meaning primarily from the context provided by sources and documentation.

The question of national identity is itself a contentious subject. As anthropologist Francesco Remotti (2012, 2017) argues, identity is often invoked despite its fluid and contested nature. It is not a fixed essence, but a narrative constructed over time, shaped by both internal evolutions and external influences. For Remotti, identity must be understood as a socio-psychological need—a desire for recognition and continuity rather than a concrete or unalterable substance. This understanding is essential when one attempts to showcase a nation like Italy through the lens of its brands: the exhibition does not aim to present a pure or unchanging essence, but rather a dynamic and evolving cultural process.

This conception of identity as constructed and relational is particularly relevant when applied to brands and companies. Often presumed to have a coherent and transparent identity, organizations instead operate within a field of multiple, sometimes conflicting, narratives. As Stuart Albert and David Whetten argued in their foundational work on organizational identity (Albert and Whetten 1985), identity is not what a company "is," but what it "claims to be"—a self-description shaped by internal perceptions, stakeholder expectations, and cultural context. This complexity becomes even more evident in legacy brands, where decades of corporate evolution, ownership changes, and shifting markets contribute to stratified and sometimes opaque identity constructs. In such cases, the notion that identity is readily accessible or self-evident is deeply problematic.

From this perspective, curatorial intervention—especially one operating externally—does not merely illustrate an existing identity but participates in its articulation. The exhibition process becomes a form of inquiry, revealing through the selection, juxtaposition, and presentation of materials the underlying values, symbols, and tensions that compose a brand's historical self. As Balmer and Greyser noted in 2003, corporate identity is always in the making, and external viewpoints can help surface latent elements that internal narratives might obscure or overlook.

Hence, historical sources in exhibitions are not mere evidentiary supports. They are narrative tools, enabling the curatorial team to frame each brand within a triadic temporal logic—past, present, future—that resists essentialism while acknowledging the need for recognition and distinction.

In this context, the artifact itself assumes a central role. Objects in exhibitions are not passive witnesses of the past: they are active mediators of meaning. Through their materiality, form, function, and aesthetic, they allow audiences to access not only the history of a product or company but also the cultural codes and values of a specific era. This is particularly relevant when dealing with design and industrial heritage, where artifacts represent the intersection of technological innovation, aesthetics, and everyday life.

Italian brands such as Algida, Bialetti, Pirelli, and Vespa embody more than market success; they are deeply embedded in the Nation's cultural memory. Their logos, product designs, and communication strategies encapsulate Italian society's aspirations, contradictions, and transformations across decades. The material culture associated with industrial production offers a privileged lens to understand economic history and shifts in social norms, taste, and identity.

Thus, showcasing these artifacts within an exhibition does more than inform; it performs an act of cultural preservation. As Smith (2006) argues in her work on heritage, the practice of exhibiting is not neutral: it is an act of meaning-making that constructs cultural memory through selection, framing, and narration. Exhibitions like *Identitalia* allow a broader audience to access and re-interpret their cultural heritage, activating a public reflection on what it means to be Italian, yesterday, today, and tomorrow.

Moreover, such initiatives are crucial when national narratives risk being flattened by globalization. They provide a space for cultural differentiation that does not fall into nationalism or essentialism but instead celebrates plurality and evolution. Clifford (1997) reminds us that heritage is not about fixed origins but about routes (transmission processes, reinterpretation, and change).

Ultimately, historical sources (physical objects, digital archives, or textual documents) are the foundation of any attempt to represent a nation's identity through exhibition. Their selection, interpretation, and presentation demand academic rigor and curatorial sensitivity. In this context, exhibitions like *Identitalia* play a pivotal role: they are reflections of the past and tools for shaping collective futures.

2 The Case of *Identitalia*: A Curatorial Overview

Identitalia. The Iconic Italian Brands is an exhibition developed to represent some of the most influential Italian brands, encompassing 113 companies across multiple sectors.

The Ministero delle Imprese e del Made in Italy—MIMIT (or Ministry of Enterprises and Made in Italy) organized the exhibition as part of a broader national initiative to enhance and promote Italian industrial culture. It was launched to celebrate the 140th anniversary of the Ufficio Italiano Brevetti e Marchi—UIBM (or Italian Patent and Trademark Office). This symbolic milestone acknowledges the pivotal role of intellectual property in constructing Italy's productive and cultural identity. The exhibition was produced in collaboration with SICAMERA, the Italian Chambers of Commerce agency, and curated jointly by the Department of Planning, Design and Technology of Architecture at Sapienza University of Rome and the Department of Design of Politecnico di Milano.

Identitalia follows the trajectory of a larger institutional project envisioned by MIMIT to narrate "Italian excellence" through the lens of design, innovation, and industrial heritage. This initiative began with the exhibition *Italia Geniale* (or Italy Genius), conceived to celebrate the creative ingenuity of Italian design. Developed by MIMIT in collaboration with ADI (Associazione per il Disegno Industriale, or Association for Industrial Design) and the ADI Design Museum, *Italia Geniale* was first presented at the Expo 2020 in Dubai, then hosted at Palazzo Piacentini, MIMIT's headquarters, in 2022. Since 2023, it has become a travelling exhibition, embarking on an international tour alongside the Amerigo Vespucci training ship, in collaboration with the Ministry of Defense and the Ministry of Foreign Affairs, making stops in cities such as Los Angeles (USA), Tokyo (Japan), Singapore, Mumbai (India), and Abu Dhabi (UAE). The exhibition served as a cultural ambassador of Italian achievements abroad, bringing the story of Italian creativity to a global audience. Within this continuum, *Identitalia* represents a second significant chapter, shifting the focus from design products to the brands that have shaped Italy's economic and cultural landscape.

A third exhibition titled *L'Italia dei Brevetti: invenzioni e innovazioni di successo* (or Italy of Patents: Successful Inventions and Innovations) has since been inaugurated, continuing the celebration of Italian ingenuity. Also promoted by MIMIT, it marks not only the 140th anniversary of UIBM, but also the 550th anniversary of the world's first patent law, issued by the Republic of Venice in 1474. Together, these exhibitions form a cohesive narrative that positions Italy's material and intellectual production at the heart of its national identity.

Inside *Identitalia*, the selection criteria were guided by the recognition of brands that contribute to the multifaceted image of Italy, reflecting what has been described as the *Four A's* of Made in Italy: *Abbigliamento* (Clothing),[1] *Agroalimentare* (Food and Beverage), *Arredamento* (Furniture), *Automobili* (Cars/Automotive), alongside other sectors such as design, sport, technology, and publishing.

The exhibition's curatorial concept was built upon a central idea: to tell each brand's story by reconstructing its trajectory across three temporal dimensions: past, present, and future. This triadic approach was not merely chronological but also interpretive. The past was explored by recovering historical images, campaigns, logos, industrial documentation, and iconic products associated with some of the brands. The present was captured through current products, branding strategies, and

[1] For a recent exhibition devoted to a fashion brand, see the essay by Migliore in this volume.

Fig. 1 Overview of the exhibition layout. Rome, Palazzo Piacentini, 2024

market positioning (Martino and Zurlo 2024). The future was evoked through open questions and emerging trajectories (Fig. 1).

In presenting the selected brands, the exhibition adopted a user-centered approach, structuring the visitor experience around a typical user journey across a full day—from the first light of morning to the dreams of the night. Brands were thus grouped into thematic sections corresponding to specific phases of daily life—from waking up in the morning till going to sleep and dreaming in the night—allowing visitors to encounter them not only as economic or cultural entities, but as part of their everyday lived experience (Micelli 2011). Each brand was then narrated through the same triadic lens within the frame of its allocated moment in the day. This experiential layout enabled a more intimate and relatable engagement with the exhibition, emphasizing how brands are interwoven with individual and collective routines, memories, and aspirations (Fig. 2).

In fact, as an initiative, *Identitalia* aimed to go beyond the boundaries of a traditional corporate showcase. It was conceived as an exhibition on the cultural and industrial identity of the nation, where brands serve as protagonists and narrators. Through artifacts, imagery, and storytelling, each brand becomes a vehicle for remembering, reflecting, and reimagining what it means to be Made in Italy.

After positioning the selected brands along the user journey, from morning to night, curators turned our attention to a more complex and nuanced question: how to narrate each brand. The curatorial challenge was to define a coherent yet flexible

Fig. 2 View of the different areas divided according to the adopted phases of the day. Rome, Palazzo Piacentini, 2024

strategy for representing brand identity within a historical exhibition format. This led to the hypothesis of showcasing what was termed "communicative artefacts": tangible and visual elements that have played, and often still play, a strategic role in shaping the brand's public image (Fig. 3).

These artefacts varied significantly depending on the company's industry, history, and communication strategy. For some, the logo served as a central visual anchor; for others, the focus was placed on the trademark, understood not merely as a legal identifier, but as a physical mark on the product, such as the engraved stamps used by the Parmigiano Reggiano cheese or the distinctive labels of the Stone Island luxury sportswear company. Curators included examples of advertising, ranging from print ads and vintage billboards to television commercials (e.g., Carosello)[2] and digital content (e.g., social media posts), as well as product packaging, which often functioned as a medium of brand expression in its own right (Fig. 4).

In some cases, the product itself was the most eloquent artefact, while in others, the company's identity was embodied through corporate architecture (as in the case of RAI—national public broadcasting company—or San Pellegrino water), or through commemorative volumes and celebratory publications. The exhibition did not attempt to impose a uniform narrative structure by assembling this constellation

[2] For Carosello, see the essay by Bosoni and Cecchi in this volume.

Fig. 3 Overview of the exhibition area dedicated to the theme of "Night." Rome, Palazzo Piacentini, 2024

Fig. 4 Overview of the typologies of artefacts used in the exhibition area dedicated to the theme of "Night." Rome, Palazzo Piacentini, 2024

of elements. Instead, it sought to capture the diversity and specificity of each brand's communication ecosystem, offering visitors a layered and multidimensional understanding of how brand identity is constructed, performed, and remembered over time.

The notion of the *Four A's* has served as a valuable heuristic to represent the breadth, diversity, and excellence that define Italian manufacturing. These sectors are economically relevant and culturally charged, embodying values such as aesthetic sensitivity, artisanal knowledge, attention to form and function, and an enduring relationship with tradition and territory. The *Four A's* are more than categories; they are archetypes of the Italian identity in global markets. They encapsulate the interplay between craftsmanship and innovation, heritage and contemporaneity, form and narrative. As Micelli (2011) has argued, the resilience and distinctiveness of the Italian industrial system lie precisely in its capacity to leverage cultural capital across multiple sectors, rooted in specific local contexts but open to global influence.

By showcasing brands across these domains, *Identitalia* offered a cross-sectional view of the Nation's economic history and cultural geography.

Brands such as Barilla (Food), Pirelli (Tires), FIAT (Automotive), Lavazza (Coffee), Alessi (Design & Homeware), and Luxottica (Eyewear) were presented as market actors and as chapters of a broader national narrative. Their histories intertwine with those of designers, engineers, artists, and entrepreneurs who have collectively shaped Italian modernity's material and symbolic infrastructure.

This narrative ambition reflects a broader theoretical positioning: exhibitions like *Identitalia* function as devices for cultural translation and nation branding. They build bridges between corporate heritage, national history, and collective imagination, reinforcing the idea that industrial products can act as cultural ambassadors.

3 Research Methodology and Brand Categorization

The exhibition's research strategy combined digital ethnography, archival exploration, and direct engagement with companies.

This dual-track methodology was structured in two core phases. The first phase consisted of an independent desk research, in which the curatorial team mapped publicly available content from each brand, extracting not only historical images and documents, but also styles of self-representation, branding discourses, and temporal logics embedded in digital narratives. The aim of this phase was to identify a stylistic and thematic baseline, useful both for curatorial consistency and for facilitating further collaboration. This exploratory step also functioned as an informational audit, revealing which brands had invested in the digital preservation and narration of their heritage, and which had neglected this dimension.

The second phase shifted from observation to interaction, involving direct company contact. Each brand was approached with a curatorial dossier that included an overview of the exhibition and a preliminary proposal based on the desk research.

As the curatorial team delved deeper, three distinct typologies of companies emerged. These categories became not only a working framework for the project, but also a critical lens through which to understand the broader state of corporate memory in Italy:

1. The Ideal Type: These companies have well-curated and accessible online archives, supported by an in-house historical office, museum, or communication team. Their commitment to heritage management facilitated a quick and fruitful collaboration. The relationship between brand identity and historical continuity was already a strategic asset in these cases. Brands in this category (often large, internationally visible firms) could provide high-quality historical documentation, professional imagery, and coherent narratives, sometimes already aligned with the exhibition's temporal structure (past–present–future). Their internal teams, often trained in archival science, communication, or brand heritage management, saw participation not as a burden but an opportunity for public engagement and reputational value. In such cases, the curatorial dialogue took the form of co-design, and the curators could act more as facilitators than reconstructors of fragmented pasts.
2. The Ambiguous Type: These companies displayed partial digital presence, with fragmented or outdated historical content. While often open to collaboration, they required additional interpretive work from the curatorial team to bridge gaps and build coherent narratives. Typically, these were medium-sized companies or brands that had undergone ownership changes, rebranding, or restructuring, disrupting the continuity of their historical identity. Internal knowledge about the company's past was often scattered among departments or stored informally (e.g., personal collections, unindexed archives). Here, the curatorial work involved a hybrid task of research and reconstruction: reassembling histories from fragments, suggesting interpretive paths, and sometimes supporting the companies themselves in rediscovering their legacy. Although more challenging, this category often yielded surprising results, as companies became aware of the value of their history through the curatorial process.
3. The Dormant Type: These brands either lacked awareness of their historical value or showed no interest in contributing to the exhibition. Occasionally, archival materials were inaccessible due to corporate transitions, lack of digitization, or cultural disinterest. This group represented the most significant curatorial challenge. Some companies had undergone acquisitions or closures, resulting in the dispersal or loss of historical documentation. Others considered their past irrelevant to their current identity or failed to recognize exhibitions as strategic opportunities. In such cases, the curatorial team had to compensate for the absence of collaboration by turning to third-party archives, design libraries, and historical publications. The work required curatorial inference, creative sourcing, and sometimes speculative reconstruction. The Dormant Type embodies the critical fragility of industrial memory when not institutionally preserved or strategically valued.

This classification highlights the methodological flexibility required in managing heritage-based projects. It also reflects broader issues in corporate archival culture, such as the lack of investment in knowledge management (Kozinets 2010) and the sporadic integration between branding, communication, and historical documentation (Davenport and Prusak 1998; Rowlinson et al. 2010) Furthermore, it underscores the gap between brand image and institutional memory. While many companies invest in marketing and storytelling, few embed historical continuity in their brand governance structures.

Such asymmetries reveal a paradox: brands are often seen as timeless cultural icons, but the infrastructure that supports their memory is frequently fragile, unstructured, or nonexistent. This insight aligns with contemporary debates on organizational forgetting and memory work (Rowlinson et al. 2010; Nora 1989; Anteby and Molnàr 2012), suggesting that memory must be actively curated to function as a resource for innovation, authenticity, and legitimacy. In this sense, the methodological experience of *Identitalia* offers not only a curatorial model but also a strategic reflection on the governance of heritage in corporate contexts.

4 Exhibition Challenges: Gaps, Access, and Cultural Sensibilities

Curating a historical exhibition like *Identitalia* entails confronting several logistical and epistemological challenges. These challenges emerged from the nature of the content—corporate histories—and the institutional, communicative, and cultural contexts in which these brands operate. Far from being mere technicalities, these issues revealed the deeper tensions at play when institutional memory intersects with national representation and public storytelling.

Among the most recurring were:

1. Disparities in Archival Culture: Despite the rich heritage of many Italian brands, archival preservation and digitization remain uneven. Some companies consider historical documentation marketing collateral, while others consider it a burden with no strategic value. The inconsistency across archival practices reflects the absence of shared standards and the variability in corporate cultures concerning the past. Historical records are not systematically preserved, indexed, or valued for many firms, particularly those without a dedicated heritage department. The curatorial team often encountered fragmented archives that were physically inaccessible or dispersed across multiple formats and sites. This lack of coherence required the team to engage in negotiations that were as much about epistemic legitimacy as logistics—persuading companies that their past was relevant and integral to the exhibition's narrative and their brand identity. As Nora (1989) noted, archives are *lieux de mémoire* only when activated; otherwise, they remain inert containers. In *Identitalia*, activating memory also means reconstructing the conditions for memory to emerge.

2. Lack of Digital Accessibility: Even when archives existed, they were often closed to the public, not digitized, or protected by restrictive communication policies. This limitation posed a dual challenge: on the one hand, it constrained the curatorial capacity to access relevant materials; on the other, it revealed a more structural issue in Italian brand culture—a limited understanding of digital openness as a form of cultural capital. In a digital age where openness fosters cultural exchange, transparency, and innovation (Giaccardi 2012) the reluctance to share or expose corporate memory suggests missed opportunities for public engagement and reputational enhancement. In this sense, the exhibition also worked as a provocation: by requesting access to memory, it stimulated reflection on the value of making memory visible. Where access was denied or partial, the team had to creatively source alternatives—from secondary literature, historical press archives, and academic publications—thus transforming curatorial limits into moments of historiographic inquiry.
3. Brand Self-Perception and Historical Discomfort: Some companies expressed reluctance in being included due to past crises, rebranding phases, or ownership changes. The desire to control brand narrative sometimes clashed with the exhibition's historical aims. In such cases, the past was seen not as a resource but as a liability. Curators were faced with the delicate task of mediating between institutional narratives and curatorial integrity, ensuring that memory was neither sanitized nor exploited. This required sensitivity to corporate storytelling's emotional and political dimensions, especially for brands that had undergone controversial phases such as labor conflicts, legal disputes, or periods of decline. As theorists of public history remind us, curating identity requires a careful balance between accuracy, inclusivity, and narrative pluralism (Grele 1991; Trouillot 2015; Ashton and Hamilton 2010). The *Identitalia* team was thus engaged in a form of applied public history, where each curatorial choice reflected an ethical stance on how to represent complexity without distortion.

The curatorial team adopted a multi-pronged and adaptive strategy to address these challenges, combining methodological rigor with creative flexibility. These strategies included:

1. Proactive mapping of available data: The team invested time in building a robust preliminary knowledge base before contacting companies. This included brand timelines, visual histories, previous exhibitions, academic references, and digital traces. This proactive mapping allowed the team to anticipate gaps, formulate informed questions, and demonstrate credibility in curatorial dialogue. It also ensured a more equitable starting point for companies of different sizes or visibility levels.
2. Collaborative dialogue with companies: The curators framed their interactions not as top-down requests but as co-design opportunities. By presenting the exhibition as a shared cultural project rather than a marketing campaign, they actively invited companies to shape their public memory. This dialogic approach helped to overcome resistance and facilitated the emergence of unexpected contribu-

tions, archival gems, unpublished materials, or personal testimonies that enriched the exhibition narrative.
3. Aesthetic standardization to balance disparities in content: Given the uneven quality and quantity of materials, the exhibition employed a unified aesthetic grammar, based on graphic consistency, typographic clarity, and modular storytelling, to ensure curatorial coherence. This standardization was not cosmetic; it functioned as a tool for equity, allowing brands with modest archives to be presented with the same dignity as larger corporations. In this way, visual design became a curatorial equalizer, helping to foreground narrative over volume.
4. Transparency in acknowledging gaps or silences: Rather than hiding the absences, the exhibition made them legible. In some cases, lacunae were explicitly signalled with editorial notes or visual markers. This transparency served two purposes: first, it maintained curatorial integrity by acknowledging the limits of representation; second, it invited visitors to reflect on the conditions of visibility and invisibility that shape public memory. As Michel-Rolph Trouillot (2015) argues, historical narratives are shaped as much by what is silenced as by what is said.

Ultimately, this approach ensured the project's feasibility and turned constraints into curatorial choices, encouraging a more critical and inclusive view of Italian brand culture. It highlighted the fact that memory is not given—it is constructed, negotiated, and performed, and exhibitions such as *Identitalia* play an active role in that construction.

5 Conclusions and Future Outlooks

The experience of curating *Identitalia* offers valuable insights for future exhibitions addressing national identity through industrial and cultural heritage. First, it underscores the need for a robust infrastructure supporting corporate archival practices, not merely as internal documentation tools, but as resources for public history and cultural diplomacy.

This need is not only cultural but strategic. Many Italian brands have begun recognizing the economic and symbolic value of their corporate heritage in recent years. Once relegated to compliance or nostalgic purposes, archival practices are now being reframed as drivers of brand positioning, identity consolidation, and stakeholder engagement. By investing in preserving and promoting their historical materials (products, advertisements, blueprints, packaging, and narratives), companies are building competitive advantages rooted in authenticity and continuity.

This strategic use of heritage serves multiple functions. Internally, it reinforces a sense of corporate culture and identity among employees; externally, it becomes a tool for storytelling, allowing brands to differentiate themselves in saturated markets. From museum-like company archives open to the public, to exhibitions, branded content, and heritage marketing campaigns, the valorization of corporate memory is increasingly seen as a way to reaffirm brand credibility, build emotional resonance with audiences, and legitimize long-term innovation. As such, heritage is no longer a passive legacy but an active asset, contributing to both economic value creation and cultural diplomacy—especially when embedded within national narratives like those explored in *Identitalia*.

Second, it reveals the symbolic power of brands as markers of identity, not because they offer fixed or pure representations of "Italianness", but because they expose its evolution, contradictions, and aspirations. As Remotti (2017) suggests, culture should be approached not as a static essence but as a process—fluid, relational, and open to contamination.

Third, the exhibition model adopted (past/present/future) effectively encouraged companies to reflect on their trajectory, fostering intergenerational narratives and future-oriented thinking. This temporal layering can serve as a blueprint for future exhibitions that bridge heritage and innovation.

Finally, the notion of identity, central to the exhibition's title and scope, must continue to be questioned and problematized. Rather than pursuing definitive answers, exhibitions should offer platforms for dialogue, critical reflection, and cultural negotiation. The curatorial challenges faced by *Identitalia* are not exceptions, but symptomatic of a broader need to rethink how we produce, share, and sustain collective memory in the twenty-first century.

Acknowledgments The authors would like to thank all those who contributed to the *Identitalia exhibition*, making it possible: Alessio Caccamo, Fabiana Candida, Laura Casaldi, Emanuele Cigliuti, Federica Dal Falco, Stefania De Angelis, Daniela Dispoto, Rita Fantini, Gilda Gallerati, Giovanni Inglese, Vincenzo Maselli, Stefano Micelli, Silvana Migliozzi, Laura Minestroni, Francesco Morgia, Alfonso Piantedosi, Luca Ruzza, Natan A. Ruzza, Zeno M. Ruzza, Alessandra Sette, Anna Turco, Mattia Francesco Ucchiello.

Competing Interests The authors have no conflicts of interest to declare that are relevant to the content of this chapter.

References

Albert S, Whetten DA (1985) Organizational identity. Res Organ Behav 7:263–295

Anteby M, Molnár V (2012) Collective memory meets organizational identity: remembering to forget in a firm's rhetorical history. Acad Manag J 55(3):515–540

Ashton P, Hamilton P (2010) History at the crossroads: Australians and the past. Halstead Press, Sydney

Balmer JMT, Greyser SA (2003) Revealing the corporation: perspectives on identity, image, reputation, corporate branding and corporate level marketing. Routledge, London

Clifford J (1997) Routes: travel and translation in the late twentieth century. Harvard University Press, Cambridge

Davenport TH, Prusak L (1998) Working knowledge: how organizations manage what they know. Harvard Business Press, Boston

Giaccardi E (ed) (2012) Heritage and social media: understanding heritage in a participatory culture. Routledge, London

Grele RJ (1991) Envelopes of sound: the art of oral history. Praeger, New York
Kavanagh G (2005) Making histories in museums. Leicester University Press, Leicester
Kozinets RV (2010) Netnography: the essential guide to qualitative social media research. Sage, London
Martino C, Zurlo F (2024) Identitalia. The iconic Italian brands. MIMIT Ministero delle Imprese e del Made in Italy, Rome, pp 16–21
Micelli S (2011) Il lavoro artigiano nella globalizzazione. Argomenti 32:53–70
Nora P (1989) Between memory and history: Les Lieux de Mémoire. Representations 26:7–24
Pearce SM (1990) Objects as meaning; or narrating the past. In: Objects of knowledge. Routledge, London, pp 125–140
Remotti F (2012) Contro l'identità. Laterza, Rome-Bari
Remotti F (2017) L'ossessione identitaria. Laterza, Rome-Bari
Rowlinson M et al (2010) Social remembering and organizational memory. Organ Stud 31(1):69–87
Smith L (2006) Uses of heritage. Routledge, London
Trouillot MR (2015) Silencing the past: power and the production of history. Beacon Press, Boston

Archives and Project: The Case of the Traveling Exhibition *Coats! Max Mara*

Ico Migliore

Abstract The essay aims to reflect on the transversality of Italian design, understood as a design process based on the encounter, in a kind of Renaissance vision of contamination between different disciplines, cultures, and contexts. At the center of the article is the case of *Coats! Max Mara*, the traveling exhibition designed by Migliore+Servetto studio which tells the story of the heritage and archive of the historic fashion company, and which has so far reached five stages (Berlin 2006, Tokyo 2007, Beijing 2009, Moscow 2011, and Seoul 2017). A project that demonstrates the deep connections in the dialectical relationship between industrial production, art, design, and communication, and where the archive proves to be a great opportunity for encounter. The essay also dedicates ample space to the link between design culture and industrial culture, defining design as the result of a continuous dialectical commitment and confrontation.

Keywords Exhibition design · Museum design · Italian design · Archives · Made in Italy

1 Introduction: The Transversality of Italian Design and the Dimension of Time

Italian design, in its deepest approach, is configured as an intrinsically transversal design process, capable of overcoming disciplinary barriers and fostering a fruitful dialogue between heterogeneous cultures and contexts. This vision of contamination, recalling a Renaissance ideal, allows design to act as a link between seemingly distant spheres, such as industrial production, art, and communication.

I. Migliore (✉)
Politecnico di Milano, Milan, Italy
e-mail: ico@miglioreservetto.com

P. Cordera (ed.), *Unfolding Made in Italy (1948-1962)*, Research for Development, https://doi.org/10.1007/978-3-032-10118-1_8

Fig. 1 The dome. *Coats! Max Mara* Exhibition. Multimedia and Graphic Design by Migliore+Servetto, Dongdaemun Design Plaza (DDP). Seoul, November 2017. (Photo by Jae Young Park)

At the center of this reflection is the traveling exhibition *Coats! Max Mara*, that I conceived and designed together with the Migliore+Servetto studio. This exhibition project aimed not only to narrate the heritage and archive of a historic Italian fashion house, but also to embody the deep connections that emerge from the dialectical relationship between different forms of knowledge and doing.

This essay aims to analyze *Coats! Max Mara* as an exemplary case of how interior design, and exhibition design in particular, has become a privileged field of experimentation, where the dimension of time plays a central role. As I previously highlighted in the volume *Time to Exhibit* (Migliore 2019), time, beyond its philosophical or scientific aspect, can take shape as a reference element in interior design, especially in temporary and permanent installations.

The research and experimentation carried out together with the Migliore+Servetto studio have led to defining the need to attribute a tangible quality to the ineffability of time, around which the project develops. In this vision, time is considered the true dimension of interior and spatial design (Fig. 1).

2 The Evolution of the Concept of *Allestimento* in the Contemporary Context

The Italian term *allestimento* (exhibition/set-up) is in itself ambiguous, semantically suspended between the meaning of "process" and "product". This uncertainty was historically accentuated by the wide range of applications, spanning from celebrations to banquets, from theater to automotive interiors, to shop windows and museums. However, this generality has dissolved with the full recognition of the discipline known as Exhibition Design. Today, the verb *allestire* (to exhibit/to set-up) unequivocally defines the tasks of qualified professionals in the field of Interior Design, who dedicate themselves to the design of temporary or permanent installations for museums, exhibitions, commercial and cultural spaces, with the primary objective of involving the public in a process of knowledge acquisition focused on the exhibited object.

The contemporary era, characterized by "liquid society" (Bauman 2002) and "risk society" (Beck 2000), has further complicated and enriched the landscape of exhibition design. The multiplication of contexts and the fluidity of boundaries between physical and virtual, original and replica, temporary and permanent, have made exhibition design a dynamic and constantly evolving field of intervention.

Despite the confusion that still characterizes the vast horizon of disciplinary applications, designers are called upon to master both the social complexity of the contemporary and that of the digital scenario. The convergence between temporary and permanent is now a given, with design experiences developing in one field and flowing into the other, drawing inspiration from mutual experimentations in terms of approach, materials, and new technologies.

The need to redefine the meaning of exhibition design is more urgent than ever today, not only due to the vast scope of global tourism towards "unrepeatable exhibitions," but also due to the growing fragmentation of users. This imposes new responsibilities on exhibition architecture, such as the conception of paths inspired not only by the objects or themes on display, but also by a kind of formative ethic. The exhibition is no longer a mere support, but a "field of experimentation" that invites designers to dare, knowing that the outcomes will almost never be definitive.

3 The Archive as Opportunity: The *Coats! Max Mara* Case

The *Coats! Max Mara* project (Seoul 2017) stands as a paradigmatic example of how a company's archive and heritage can be transformed into an engaging and innovative exhibition narrative.[1] The exhibition, already presented in several international cities such as Berlin (2006), Tokyo (2007), Beijing (2009), Moscow (2011),

[1] Established in 1951 in Reggio Emilia by Achille Maramotti, Max Mara is a world-renowned Italian fashion house. At the time, Maramotti had a pioneering vision: to apply the principles of industrial production to high-quality women's clothing, a revolutionary idea in Italy, where fashion was still an artisanal craft. The company quickly became celebrated for its impeccable tailoring and iconic outerwear, most notably the legendary 101,801 coat, a timeless piece that remains a bestseller today. Max Mara is a symbol of sophistication, quality, and timeless elegance. Its success is built on a legacy of consistency and a continuous dialogue between classic design and modern sensibility. As a global leader in Italian prêt-à-porter, the company has expanded to include numerous lines and maintains a strong presence around the world. The company's vast archive and rich heritage serve as a testament to its long history of innovation, making it a powerful subject for an exhibition that explores the evolution of corporate culture and design.

arrived in Seoul in 2017 at the Dongdaemun Design Plaza (DDP)—one of the most important and iconic landmarks of the South Korean capital, designed by Zaha Hadid in 2014.

With the Migliore+Servetto studio, we conceived a majestic architectural structure, inspired by the utopian monoliths of Étienne-Louis Boullée or the Cenotaph he imagined for Sir Isaac Newton[2] and the Pantheon.[3] At the center of Art Hall 1 (approximately 3000 square meters) of DDP an imposing dome supported by seven irregular volumes emerges from the darkness, shaped by plays of light and shadow, in dialogue with the surrounding architecture. This structure was conceived as an open and extremely permeable space, which invited the visitor to create personal and multiple paths. Beneath the dome, a vast vault dematerialized like a large planisphere to create a dynamic and pulsating narrative, conveying Max Mara's sensitivity towards heritage and art. The young artist Yiyun Kang interpreted the theme with a digital performance that envelops the observer in a poetic story of fabrics, volumes, bodies, and Italian architectures.

The exhibition was structured into seven *Wunderkammern*[4] or chambers of wonders, which delineated and punctuated the exhibition path, exploring the brand's evolution from the 1950s to today. The concept of *Wunderkammer*, historically understood as a place for collecting and displaying rarities and works of art, is here reinterpreted to describe installations that develop a legitimate and effective way of designing, with excellence in the compositional harmony of individual elements. Each *Wunderkammer* was like a precious casket that opened around the exhibited objects, serving as a guardian and megaphone of corporate memory and identity. Crucially, each of these chambers boasted a distinct title and theme, guiding visitors

---

[2] This celebrated unbuilt project, conceived by the French architect Étienne-Louis Boullée in 1784, pays homage to Sir Isaac Newton. The work, never realized, stands as a quintessential example of visionary Neoclassical architecture, characterized by elementary geometric forms and a monumental scale. The cenotaph was imagined as a vast, hollow sphere. By day, it would replicate the celestial vault through apertures that filtered light, and by night, it would simulate daylight with a luminous globe suspended at its center. This project reflects Boullée's profound admiration for science and his pursuit of the sublime in architecture, deeply influencing subsequent architectural thought.

[3] The Pantheon in Rome, built by Hadrian (c. 118–128 CE), is a masterpiece of Roman architecture. Its iconic concrete dome, the world's largest unreinforced concrete dome, culminates in a central oculus. Originally a "temple to all gods," it was converted into a church in 609 CE. It stands as a symbol of advanced engineering and geometric harmony (its internal height, measured from the floor to the apex of the oculus, is precisely equal to the diameter of its rotunda). Its remarkable preservation and enduring influence make it a foundational monument in architectural history.

[4] A *Wunderkammer*, or cabinet of curiosities, was a collection room prevalent in Europe from the sixteenth to eighteenth centuries. These pre-modern museums housed diverse objects: *naturalia* (natural specimens), *artificialia* (man-made artifacts, including art), *scientifica* (scientific instruments), and *exotica* (items from distant lands). They served as microcosms, reflecting a collector's attempt to encompass all knowledge and the world's wonders, blending art, science, and ethnography before the advent of modern museum classifications.

Fig. 2 *The Creative Studio. The Democratic Coat, Designers, and Pop* (1960s), *Coats! Max Mara*, Exhibition, Multimedia and Graphic Design by Migliore+Servetto, Dongdaemun Design Plaza (DDP). Seoul, November 2017. (Photo by Migliore+Servetto)

through different facets of the brand's evolution. These themes included *The Founder. Achille Maramotti and the Dream of the Coat* (1950s); *The Creative Studio. The Democratic Coat, Designers, and Pop* (1960s); *Colorama. The Colored Coat and the Art of Experimentation* (1970s); *The Icon. The Magic of the Coat and Italian Craftsmanship* (1980s); *The Photo Set. The Coat's Narrative and the Perfect Image* (1990s); *The Max Mara Women. The Coat's Journey and the Feminine Universe* (2000s), and *The Fashion Show. The Runway Between Backstage and New Trends* (Early 2010s) (Fig. 2).

Offering a specific scenario for each decade, these chambers were enriched by a sophisticated ensemble of sounds, music, scents, and tactile experiences, which completed the narrative, creating unique microcosms with a fully immersive atmosphere. The archive, the place of memory preservation par excellence, was a necessary phase of the design work, fundamental for building a coherent story with the objects on display and establishing a common identity for them. The Seoul exhibition aimed to reach the heart of Max Mara's heritage, revealing its strong artistic scope and the continuous innovation that has always characterized this iconic Made in Italy brand (Fig. 3).

Fig. 3 *The Creative Studio. The Democratic Coat, Designers, and Pop* (1960s), *Coats! Max Mara* Exhibition, Multimedia and Graphic Design by Migliore+Servetto, Dongdaemun Design Plaza (DDP). Seoul, November 2017. (Photo by Migliore+Servetto)

4 Design as a Dialectical Commitment: Industrial Production, Art, and Communication

The *Coats! Max Mara* project illustrates how design is the result of a continuous dialectical commitment and confrontation, especially in the relationship between design culture and industrial culture. The exhibition highlights how contemporary commercial spaces have evolved from simple points of sale to complex environments for product communication and representation, where tools and devices typical of cultural or artistic installations converge in the construction of narratives. Contemporary cultural spaces, as well as commercial ones, are, in fact, or should be, places of high expressive synthesis, capable of giving centrality not only to the product but also to the project that generated it, the entrepreneurial idea, the cultural context that inspired it, and the message it intends to communicate.

This translates into the discovery of qualities made of traditions, entrepreneurial responsibilities, and interpretations of contemporaneity. The work of Carlo Scarpa,[5] such as the famous Olivetti showroom in Piazza San Marco in Venice, provides an emblematic example of how the exhibition element persists in retail, uniting the modernity of design with traditional local techniques and materials to create exhibition structures that, together with light, define the dynamic rhythm of the exhibition and seem to expand the walls into a narrative.

Andrea Branzi (2002), reflecting on the "liquid society," observes how individual freedom coincides with the liquefaction of the system, nullifying the distances between private and collective interests, and between the design of permanent and provisional structures. In this context, exhibition design is called upon to translate the idea of perennial mutation and to confront the multiplication of contexts, while maintaining the link with the genius loci and the historical matrix of the places.

The contemporary project, although more fluid, does not evade the need to imagine itself in relation to the past time that represents the history of places. The designer, in this complex scenario, assumes the role of director and privileged mediator. As designer Enzo Mari (1991) argued, the responsibility for setting up the exhibition space should fall to the architect, who must interpret the brand's identity without distorting their own vision, respecting cultural and economic-political reasons. The exhibition, from a simple scenography, then evolves into a "new concept of script" narrated through behaviors, tools, and structures. In this intertwining, the relationship with the world of communication has become a pivot of innovation, involving technologies, multimedia, virtuality, ecology, and sensoriality in defining the exhibition context. Architect and designer Achille Castiglioni, for example, emphasized the importance of making every project memorable,[6] creating artificial landscapes where the exhibition project becomes a driver of behaviors and thoughts, giving life to a complex three-dimensional script.

---

[5] Carlo Scarpa (1906–1978) wasn't just an Italian architect and designer; he was a master of sensory experience, deeply rooted in Venetian culture and its rich tradition of craftsmanship. His work stood out for its almost obsessive attention to detail, the way he brought together diverse materials, and his brilliant handling of space, often blurring the lines between building and sculpture. His Olivetti showroom in Venice (1957–1958), situated in Piazza San Marco, exemplifies his unique genius. Scarpa transformed a constricted space into a fluid, luminous environment. Rather than a mere display area, he crafted a sensory experience, employing varied marbles, brass, and wood with exquisite precision. Key features include the seemingly suspended staircase and the artful manipulation of light and shadow, demonstrating his ability to infuse modern interventions with a profound respect for historical context and traditional craftsmanship, transcending mere functional design.

[6] During the years, Mara Servetto and I worked as assistants to Achille Castiglioni at Politecnico di Milano—having both earned our Architecture degrees with him from Politecnico di Torino—Castiglioni frequently employed the adjective *unforgettable* during lectures and the painstaking hours we spent developing them together. He used it to underscore how the essential and foundational quality of an exhibition design project was its ability to forge a powerful, identifying synthesis for the space. This synthesis, stemming from the core concept (which he termed the primary design component), allowed the exhibition to become etched in the visitor's memory, conveying a message and prompting a series of reflections within them.

5 *Coats! Max Mara* as a Narrative and Polyphonic Experience

The traveling exhibition *Coats! Max Mara* fully embodies the vision of exhibition design that evolves into a narrative space, capable of involving a heterogeneous audience through variable and multiple paths, both physical and immaterial. The goal is to transform the visitor's simple advancement towards content into a secular procession, which guarantees a sense of anticipation, preparation, and almost devotion. This concept is based on a cognitive evolution and a stratification of content and apparatus, stimulating and satisfying the visitor's curiosity (Fig. 4).

The seven *Wunderkammern* of *Coats! Max Mara* offer a "slow-release" immersive experience. These slow-release projects serve as activating moments for subsequent personal research on the content and container, leading to the birth of a new aesthetic of actions and usage behaviors through new interfaces of physical interaction with the visitor. The complexity and richness of the content are translated into a system of micro-museums within the museum, where the visitor can choose their own thematic path and delve into details without losing the overall vision.

This approach moves away from the "nail-on-the-wall museum," which, as I have previously illustrated in the volume *Time to Exhibit* (Migliore 2019), merely

Fig. 4 Interiors of the Dome, with the digital installation of Yiyun Kang. *Coats! Max Mara* Exhibition, Multimedia and Graphic Design by Migliore+Servetto, Dongdaemun Design Plaza (DDP). Seoul, November 2017. (Photo by Jae Young Park)

displays works in static showcases without a critical order, and from the "amusement park museum" (Migliore 2019), which aims for ephemeral spectacularity and superficial emotional impact, neglecting the depth of content. What we define as "narrating space"—of which *Coats! Max Mara* is an example—stands as a "third design path" a "polyphonic space for culture."

Here, the visitor is the active protagonist, free to move and personalize their own story among the events woven into the exhibition fabric. Design is not limited to exhibition structures, but creates relationships: between space and objects, between objects and people, between space and people. Experiments with new technologies, light, dynamics, and texts contribute to defining the rhythms and times of the overall "spatial editing," which, borrowed from cinema, transforms the exhibition into a sum of narrative sequences and engagement.

The "mirroring effect"[7] is a key concept in this approach, where the exhibition allows the visitor to reflect themselves in it and feel part of the narrative. This implies that the space and its contents are returned to the diversity of users, encouraging real relationships and activating deep social dynamics. Pause, stop, and silent contemplation are considered design stages for the cultural growth of the visitor. The effectiveness of a narrating museum is therefore measured, as previously highlighted, by its ability to produce "slow-release effects," stimulating curiosity and interest that will prompt the visitor to further personal exploration, in other places and times.

6 From Hidden Archive to Dynamic Cultural Seed

Within the vision I explicitly detailed in 2024 for the future of cultural spaces (Migliore and Servetto 2024), archives assume a fundamental role that goes far beyond simple preservation, transforming into dynamic catalysts of culture and narration.

The "Museum Seed" concept proposes that museums and cultural venues, like seeds, grow, transform, and expand into an "augmented" and constantly evolving version. In this context, works from the archives will "go on stage" thanks to new technologies, telling their stories on the special stage that is the museum. The museum is no longer just a protective showcase but a stage where the narrative element combines with the conservative element to bring the visitor "inside a scene dominated by the artwork" through empathy and emotions.

In our eight-point manifesto, written along with Mara Servetto, I had the opportunity to explain how the museum can also be described in relation to archives (the

[7] The mirroring effect in psychology refers to the unconscious imitation of another person's gestures, speech patterns, or expressions during social interaction. This automatic mimicry, often linked to "mirror neurons," primarily serves to build rapport, empathy, and social connection. When individuals subtly mirror each other, it fosters a sense of understanding and liking, crucial for harmonious interactions. See Chartrand and Bargh (1999) and Iacoboni (2009).

topic is covered in the chapter From Hidden Archive to Dynamic Cultural Seed): presenting it as a promoter of corporate culture, a builder of brand identity, and, consequently, a witness to social and cultural evolution.

The sources highlight that increasingly, large industrial groups develop their archives to create true collections that, from private, become accessible to the general public, both occasionally and permanently. These collections not only refer to the brand's identity but also bear witness to social evolution and play a growing role in the production of cultural content. Corporate archives are capable of generating stories that go beyond the individual entrepreneurial venture, building multiple narratives that offer insights into a given era and corporate culture. Company museums offer a space for a narrative that weaves together tradition, experience, and innovation, showing how all fields of entrepreneurial culture are permeated by ingenuity, ideas, and intuition. Many international companies understand the importance of preserving their archives and the memory of their history, not only as historical-cultural heritage, but as a living and rich testimony that forms their productive, corporate, and cultural identity. These companies seek to reinterpret and attract new generations of consumers through the presentation of their heritage.

The challenge is not only to preserve archives for their historical and commercial value, but also to make them alive, active, and profitable, communicating their richness in a direct and original way.[8] In this process, the developments in Digital Humanities are providing significant help. Jeffrey Schnapp (2011) coined the term "knowledge design" to define new models of representation and narration of historical memories made possible by new technologies. This approach allows for the creation of interactive experiences in museums and the experimentation of new modes of communicative design. "Knowledge design" uses databases and machine learning tools to create an engaging and unprecedented experience, with interactive visualization and storytelling narrative processes.

However, despite the communicative power of new digital media, the creativity of the story and its staging always requires an "exhibition/set-up direction" that guarantees the spatial quality in which the visitor is welcomed for their cognitive journey into history and archives. These archives are considered repositories of ideas, inspirations, and visions that can still be stimulating and illuminating if communicated effectively.

This holistic vision of contemporary design, which integrates graphics, interiors, architecture, and multimedia, represents a profound evolution of the Renaissance idea of design. As is well known, in the Renaissance, figures like Leonardo da Vinci or Michelangelo embodied the ideal of the artist-engineer, capable of conceiving works that fused art, science, and technique. Design was conceived in a unitary way: a fresco was an integral part of the architecture that housed it, and sculpture dialogued with the surrounding space, often under the direction of a single brilliant mind that oversaw all disciplines. However, that synthesis was predominantly

[8] On this, see the essays in Part II, *Archives for Memory and for the Future*, for examples of new approaches to archives online and onsite.

two-dimensional or static three-dimensional and limited by the available tools and technologies.

Today, the advent of digital and multimedia has expanded the concept of "total project" to a dynamic, interactive, and sensory dimension, allowing for the manipulation of not only forms and spaces, but also light, sound, and information, in ways that Renaissance masters could only have imagined, pushing the concept of integration to unprecedented levels of complexity and immersion. It is indeed necessary that contemporary design be increasingly integrated between graphics, interiors, architecture, and multimedia to create cohesive and immersive spatial and communicative experiences.

Today, design can no longer afford to operate in silos. Whether it's a building, an exhibition space, or a product, the success of a project lies in its ability to engage all senses and offer a complete user experience. Graphics are not just an ornament, but an element that guides perception, provides information, and strengthens the identity of the place. Architecture defines the structure and container, while interior design shapes livability and atmosphere. The crucial element that ties everything together is multimedia: interactive screens, projections, sounds, and dynamic lighting systems are no longer an option, but essential tools to inform, entertain, and connect people with the space, making the experience fluid and memorable. This synergy is fundamental for a significant and lasting impact.

In this scenario, archives are like essential components of a museum that is a "living organism," not just a container, but a producer of content that adapts to changes and interacts with different audiences. Their ability to generate dynamic narratives, supported by technology and innovative spatial design, is crucial for the future of cultural venues.

7 Conclusions: The Archive as an Encounter in a Polyphonic Perspective

The case of the *Coats! Max Mara* exhibition by Migliore+Servetto offers a clear demonstration of the transversality of Italian design and its ability to synergistically integrate industrial heritage, art, design, and communication.

The project reveals how the archive, far from being a mere repository of artifacts, transforms into a great opportunity for encounter. The *Wunderkammer*, the digital projections by Yiyun Kang, and the attention to sensoriality are not just scenographic artifices, but devices that activate memory and invite a deep dialogue with the history and identity of the Max Mara brand. Exhibition design, in this vision, redefines the scenography project as memory and active behavior within a dense script. The true protagonist is no longer just the object of the exhibition, but the totality of the cognitive experience and the dynamic and empathetic relationship between the different actors—objects or subjects—of the exhibition.

The polyphonic project aims not so much at what is built, but at the creation of magnetic fields of attraction or repulsion, designed to define the right time and the right distance between things. In a polyphonic exhibition, each element possesses its own qualities and can be the subject of a story that unfolds for a vast and varied audience. The completeness and meaning of the parts emerge only from the whole and from the overall definition of the quality of the space, which thus becomes a place.

The architecture of communication, therefore, shapes not only museums understood as buildings for the preservation of culture, but cultural plazas with an open character, where the flow of individuals can interact, rediscovering physical sharing as a founding element of culture itself. Ultimately, it is in the encounter with the public and within a community that a cultural object defines itself as such, not in isolation. The narrating museum, as a paradigm of polyphonic space, completes the reason for its own exhibition, configuring itself as radical innovation that is transforming the structures and modes of cultural procurement for each of us today, while simultaneously laying the foundations for the future of exhibition space.

The case of *Coats! Max Mara* fits perfectly into this perspective, with the specific objective of showing the vitality of Italian design in creating experiences that go beyond mere display, to become true moments of connection and cultural enrichment.

Competing Interests The author has no conflicts of interest to declare that are relevant to the content of this chapter.

References

Bauman Z (2002) Modernità liquida. Laterza, Rome-Bari

Beck U (2000) La società del rischio. Verso una seconda modernità. Carocci, Rome

Branzi A (2002) L'allestimento come metafora di una nuova modernità (Exhibition design as metaphor of a new modernity). Lotus Int 115(98):101

Chartrand TL, Bargh JA (1999) The chameleon effect: the perception-behavior link and social contagion. J Pers Soc Psychol 76(6):893–910

Iacoboni M (2009) Mirroring people: the new science of how we connect with others. Farrar, Straus and Giroux, New York

Mari E (1991) Per una qualità dell'allestire. Progex 7:6

Migliore I (2019) Time to exhibit. Directing spatial design and new narrative pathways. Franco Angeli, Milan

Migliore I, Servetto M (2024) Museum seed. The futurability of cultural places. Electa, Milan

Schnapp J (2011) Knowledge design v_1.0. Unpublished draft for presentation at *emerging disciplines II conference*. Rice University. http://jeffreyschnapp.com/wp-content/uploads/2011/09/Knowledge-Design.pdf. Accessed 20 Aug 2025

Digital Technologies to Promote and Valorise Design Culture

Mauro Ceconello and Davide Spallazzo

Abstract Industrial design, seen as cultural heritage, reflects both material production and creative thought of its era, yet often remains underexplored. In the Milan area—birthplace of iconic Italian design—former studios still preserve the spirit of the golden age. Though inactive, these sites remain tied to the processes and ideas that shaped them. Design research plays a key role in unlocking archives to build narratives, inspire new interpretations, and foster experiential learning. This chapter presents two cases of digital heritage enhancement: the digitisation of Giovanni Sacchi's archive, connecting sketches, drawings, photos, and prototypes to reconstruct design trajectories; and Looking for Achille Castiglioni, a mobile app offering a geolocated tour of Castiglioni's works in Milan, recontextualising them in their original sites. These examples show how digital tools can preserve, interpret, and share design culture, turning archival traces into accessible, dynamic experiences.

Keywords Design culture · Digital archives · Digitisation · Cataloguing · Location-based mobile experiences

1 Designer's Studios and Archives: An Outstanding Legacy

Following the Second World War, Italian industrial design began to attain international recognition, thanks to the vision and initiative of pioneering entrepreneurs who entrusted talented designers with the development of their products. Many objects that were once part of everyday life have since been canonised as design icons and are now exhibited in major museums. However, the cultural heritage

M. Ceconello (✉) · D. Spallazzo
Politecnico di Milano, Milan, Italy
e-mail: mauro.ceconello@polimi.it

P. Cordera (ed.), *Unfolding Made in Italy (1948-1962)*, Research for Development, https://doi.org/10.1007/978-3-032-10118-1_9

associated with this phenomenon is often reduced to the object itself[1]—the final artefact—while the broader creative process and the disciplinary expertise of the Design Masters remain largely underrepresented. A more comprehensive understanding of Design as cultural heritage must also account for these intangible dimensions, including the epistemic, procedural, and contextual elements embedded within the act of designing.

Design, when recognised as a form of cultural heritage, not only communicates the tangible qualities of the artefact but also conveys the underlying system of values from which it emerged. While the exhibition of industrial products offers a valuable entry point for promoting design culture, it must be accompanied by appropriate contextualisation to capture the complexity of the design process. Public institutions and private foundations have contributed significantly to the construction of a design culture in Italy by organising exhibitions, curatorial initiatives, and critical debates, as well as by providing spaces that serve as both platforms for dissemination and places for the valorisation of Italian design—what might be described as *Modo Italiano* (the Italian way).

Notable among these are the Design Museum of the Milan Triennale, which functions simultaneously as an exhibition venue, a tourist attraction, and an educational hub for aspiring designers, and the recently inaugurated ADI Design Museum Compasso d'Oro. In addition to these institutions, the city of Milan hosts several studio-museums and archival foundations—such as those dedicated to architects and designers Vico Magistretti, Franco Albini, Achille Castiglioni, and the Giovanni Sacchi Archive—which preserve a wide array of documentation, prototypes, models, and drawings. These materials afford critical insights into the historical development of specific design practices, enabling detailed reconstructions of creative trajectories.

The debate surrounding the valorisation of this heritage increasingly intersects with discussions about digital innovation and cross-media engagement, as technological developments continue to generate new modes of cultural participation. The adoption of bespoke digital applications—whether accessed remotely or *in situ*—has enabled museums and cultural institutions to broaden their audiences and diversify their communicative strategies. In this regard, the concept of ubiquitous computing (Weiser 1991; Greenfield 2006) proves particularly relevant, as it highlights the expanding presence of information and communication technologies in everyday life, and their capacity to shape new experiential formats for cultural consumption.

An illustrative case is the international online forum *Narrare con l'archivio* (or Narrating with the Archive), organised by the Vico Magistretti Archive in November 2020 during the COVID-19 pandemic (Pavoni and Riccini 2020). The initiative sought to explore the potential of archives as generative tools for historical narration, interdisciplinary research, and pedagogical innovation. Bringing together

[1]On Italian design displays shaped by collectors' taste, see Costa and Di Pietro's essay on the Cerulli Foundation in this volume.

scholars and practitioners from diverse institutional backgrounds, the event facilitated a transdisciplinary dialogue and resulted in a corpus of shared reflections on the role of archival material in design culture.[2]

It is within this framework that another relevant initiative can be situated: the research project D.E.SY.—Designing Enhancement Strategies and Exhibit Systems for the Italian House Museums and Studios, developed at the Design Department of Politecnico di Milano. The project aimed to elaborate design-driven strategies for the valorisation of *faber* spaces—studio-museums and house museums associated with artists and designers, active between the late nineteenth and early twenty-first centuries. These environments are not only the physical settings in which creative activity was performed but also constitute forms of spatial testimony—repositories of intangible and material heritage that reflect the everyday practices and intellectual trajectories of their former occupants.[3]

Such sites are increasingly acknowledged by heritage organisations, including the International Council of Museums (ICOM), as significant components of cultural patrimony. Whether under public or private ownership, these house museums, ateliers, and studios can be considered museums in their own right. They continue to embody the narrative of their original function, even when this narrative remains latent or partially obscured. Through visual documentation, architectural features, and material traces, these spaces reveal complex relationships between identity, memory, and creative production. Furthermore, the stewardship of these places—often undertaken by heirs, foundations, or dedicated curators—underscores the interplay between personal histories, biographical narratives, and the cultural processes of making, thereby offering fertile ground for further investigation into the intersection of design practice and heritage discourse.

2 Castiglioni and Sacchi: *Faber* of the Italian Design

Among the numerous testimonies and case studies analysed, two stand out as particularly emblematic of the so-called *golden years* of Italian design: the Fondazione Achille Castiglioni—an exceptional repository of creativity, encompassing an extensive collection of drawings, prototypes, and inspirational artefacts—and the models' archive of Giovanni Sacchi. Castiglioni and Sacchi may be regarded as exemplary figures of the *faber* tradition. Throughout their careers, they shaped improved visions of reality through design, drawing, and modelling, addressing essential concerns such as functionality, industrial optimisation, and communicative efficacy. Though operating in distinct yet complementary roles, both made

[2] For a comparable experience during the Covid-19 pandemic, see the essay by Feraboli in this volume.

[3] For unconventional sites of display, see Bollati and Collina's essay in this volume.

substantial contributions to the development and international standing of Italian industrial design.

Achille Castiglioni, trained as both an architect and a designer, was profoundly influenced by rationalist principles. In a well-known interview with Piero Polato, he distanced himself from stylistic affiliations, instead advocating for a design approach rooted in critical self-questioning and functional necessity. He maintained that design decisions required objective justification—such as ease of production, usability, or the capacity to evoke specific behaviours or meanings—while purely subjective motivations should be avoided (Polato 1991). Polano (2006) similarly notes Castiglioni's fascination with the anonymous design of everyday objects. A significant portion of his work exemplifies a practice in which the subjective act of designing aligns with the implicit rationality of vernacular artefacts—a phenomenon Polano terms "undesign". Castiglioni famously asserted that "design demands observation" (Antonelli and Guarnaccia 2000). From this epistemological stance, his iterative design process unfolded through drawing and modelling, employing materials such as plasticine and plaster. Prototypes, in particular, served as critical tools for engaging with and testing the three-dimensionality of his designs. In this context, his long-standing collaboration with Giovanni Sacchi proved to be pivotal.

Unlike Castiglioni, Sacchi did not begin his career in the design field. Trained as a foundry model maker during his youth, he initially operated a mechanical model shop, even in the face of postwar adversity. A turning point occurred when he encountered designer Marcello Nizzoli (Polato 1991) which led him to transition from precision-engineered mechanical models to experimental and representational prototypes for architects and designers. These new artefacts—designed for study, verification, and presentation—enabled practitioners to assess the strengths and weaknesses of their projects and to communicate their visions more effectively to stakeholders (Maldonado 1987; Salvarani 1992; Consalez and Bertazzoni 1998). In doing so, Sacchi offered an invaluable contribution to Italian industrial design, cultivating a collaborative craft practice that supported the design process from conceptualisation to final presentation.

His role extended well beyond that of an executor: Sacchi actively identified and addressed design flaws, and developed a deep, intuitive grasp of the design language—often anticipating and articulating ideas not fully expressed in drawings (Polato 1991). The final models produced in his workshop were so refined in detail that they were frequently used in commercial campaigns. Recognising his lifelong contribution, Sacchi was awarded the Compasso d'Oro for career achievement in 1998, supported by many leading designers of the time.

Both Castiglioni and Sacchi enjoyed prolific careers, producing a vast body of work that continues to reflect their distinctive approaches and expertise. Much of this material remains preserved in their original studios and workshops, now transformed into archival institutions of exceptional cultural value. A first step toward public accessibility occurred in 2006, when the heirs of Castiglioni signed a 5-year agreement with the Milan Triennale, resulting in the opening of the Studio Museo Achille Castiglioni as an official archive. In 2011, the Fondazione Achille Castiglioni was formally established, with the principal objective of cataloguing, preserving, and disseminating the designer's legacy—comprising drawings, photographs, models, films, objects, books, and periodicals.[4]

Likewise, the Archivio Giovanni Sacchi was inaugurated in 2009 through a collaboration between the Municipality of Sesto San Giovanni, the Fondazione ISEC (Istituto per la Storia dell'Età Contemporanea), and the Fondazione Cariplo. This institution collects and preserves models, products, technical drawings, photographic documentation, machinery, and other equipment from Sacchi's original workshop.[5]

However, physical visits to these archives, while valuable, are insufficient to ensure widespread dissemination and engagement with their contents. Barriers remain in terms of the materiality of the artefacts, the formats of documentation, and the accessibility of the physical sites.

To address these limitations and to enhance the cultural value of these institutions, two pilot projects have been developed. The first focuses on the digitisation and cataloguing of archival material—such as documents, drawings, models, and prototypes—associated with selected iconic designs, thereby enabling broader access and interpretive engagement. The second explores how archival content related to Castiglioni's work, preserved in his Milanese studio, can be accessed through a geolocated mobile application. This approach aims to create a situated and interactive user experience, leveraging digital tools to make design heritage more legible and participatory within the urban context.

3 Reconstructing Design Processes Through Digitisation: The Sacchi Case

The Milan Triennale conserves a significant portion of Giovanni Sacchi's legacy: a collection of 312 wooden models currently held by the Lombardy Region. As part of a broader initiative to promote access to this cultural asset, the Region's cultural department entrusted the Design Department of the Politecnico di Milano with a digitisation project aimed at enhancing public accessibility to selected works. The initial concept focused on performing high-quality 3D scans of the models, with the goal of producing digital twins through advanced manufacturing technologies. This would allow for both the preservation of the originals and their recontextualised exhibition. However, the scope of the project gradually expanded to encompass a broader objective: reconstructing and communicating the design processes underlying the creation of these artefacts.

Indeed, in addition to the memories of collaborators and contemporaries, the design processes of the Masters of Italian design are also traceable through a variety

[4] Fondazione Achille Castiglioni. Available via https://www.fondazioneachillecastiglioni.it/la-fondazione. Accessed 20 August 2025.

[5] Archivio Sacchi, Available via https://www.archiviosacchi.it/.

of artefacts—sketches, drawings, preparatory drafts, models, and prototypes—which collectively document the various stages of product development. These materials constitute a valuable didactic resource for future generations of designers and offer a lens into the iterative logic of industrial design practice. For this reason, the project adopted a comprehensive approach, aimed at reconstructing entire design trajectories through systematic digitisation.

From among the vast body of work developed by Sacchi, a selection of products was made based on the availability of relevant documentation preserved in design studios and corporate archives. Priority was given to projects for which complete sets of materials were available—from initial concept sketches and inspirational resources to working drawings, intermediate models, and final prototypes. To test the methodological soundness and relevance of this approach, six of Sacchi's models were selected based on their significance in the canon of Italian design and the availability of detailed archival material. These include: the Gibigiana lamp (1980, A. Castiglioni for Flos); the Tama lamp (1970–1977, I. Hosoe for Valenti); the 4870 chair (1984, A.C. Ferrieri for Kartell); the 4822/44 stool (1977–1979, A.C. Ferrieri for Kartell); the Trattopen marker (1975–1976, Design Group Italia for Fila); and the Rialto telephone (1975–1976, Design Group Italia for Siemens).

The project aimed to establish a replicable methodology for digitisation, applicable to a heterogeneous set of artefacts from diverse sources and associated with different design processes and production outcomes. In recent years, digitisation has become a widely adopted practice, facilitated by technological advancements that have transformed the ways in which data is generated, accessed, stored, and disseminated. Digital libraries now complement traditional archival infrastructures, offering ubiquitous access via the internet and mobile platforms. A prominent example is Google Arts & Culture, which employs Street View technology to provide virtual access to museums around the world and offers curated thematic content.

Yet, for cultural institutions, the real challenges lie in the cataloguing and dissemination of digital data across multiple platforms and for diverse audiences—including both specialist researchers and the general public. Numerous European initiatives have addressed these issues, including projects such as Minerva, Michael, Michael Plus, Athena (Angelaki et al. 2010) and the Europeana platform.[6] In Italy, the Istituto Centrale per il Catalogo e la Documentazione (ICCD) has established formal standards for the digitisation of visual materials (Auer et al. 1998) and for the proper cataloguing of associated metadata. In this context, the Lombardy Region developed SIRBeC (Sistema Informativo Regionale dei Beni Culturali), a regional cultural heritage database and cataloguing system (Degiarde 2007) which served as the reference framework for the project.

The digitisation workflow was articulated into three main phases—retrieval, digitisation, and cataloguing—each adhering to relevant legal and technical standards.

[6] Available via https://www.europeana.eu/. Accessed 20 August 2025.

1. Retrieval. Archival materials were identified and collected from the designers' archives and from the firms involved in the production of the selected objects. The primary objective was to gather documents and artefacts capable of reconstructing the design thinking behind the final product. Consequently, commercial and marketing materials were excluded from the scope of the study.
2. Digitisation. The digitisation phase was calibrated according to the nature and dimensions of the materials. Large-format drawings (e.g., technical drawings) were captured through digital photography; sketches, documents, slides, and transparencies up to A3 size were scanned using high-resolution equipment; Sacchi's wooden models were digitised through 3D laser scanning; and one key prototype was documented through 360° photographic capture. The entire process followed the technical specifications outlined by ICCD (Auer et al. 1998) achieving the highest digitisation standard (Level A) for large-format and conservation-grade reproductions. For accurate colour calibration, all materials were digitised with Kodak grey scale and colour targets (Fig. 1).

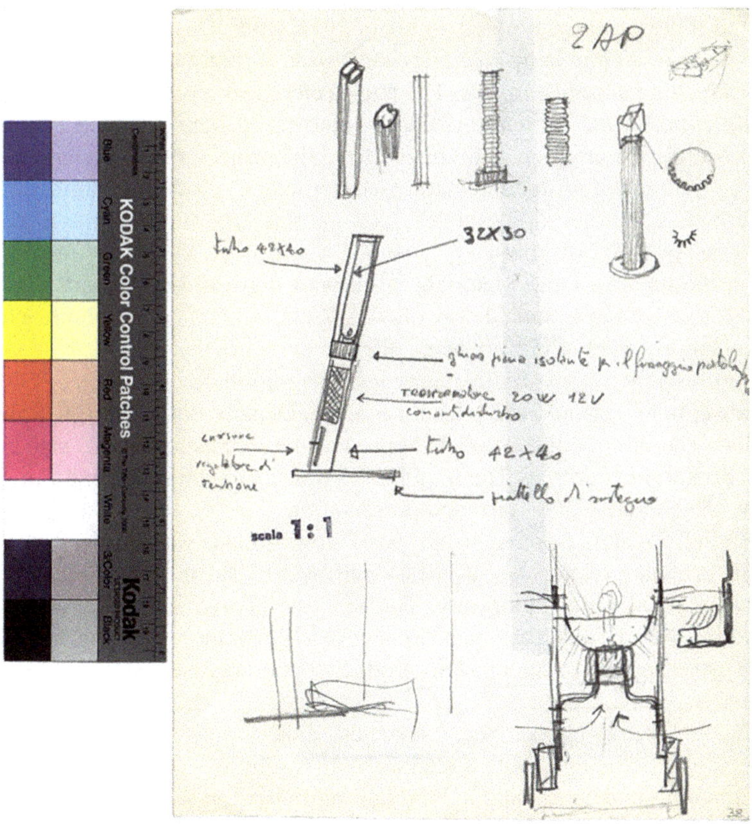

Fig. 1 Study sketches for the Gibigiana Lamp, digitised with Kodak target. (Image courtesy of Fondazione Achille Castiglioni—SIRBEC)

3. Cataloguing. The digitised materials were uploaded into the SIRBeC database in compliance with the classification protocols established by Regione Lombardia. Specifically, all drawings, intermediate models, and final prototypes were catalogued as works of art (Opera d'Arte, record type OA), thereby acknowledging their cultural and artistic significance. Photographs, slides, and transparencies were catalogued using the photographic record type (F). Furthermore, a dedicated record type for industrial design (DES R.L.) was employed as a test bench to connect and contextualise the various components of each project. This record describes the overarching design concept, tracks subsequent iterations or re-editions, and provides authorial and contextual metadata.

4 Gibigiana Lamp: Storytelling Through Archives Resources

The digitisation activities undertaken to investigate Achille Castiglioni's design process are exemplified by the case study of the Gibigiana lamp—one of the most iconic and representative artefacts in his body of work. Designed in 1980 for Flos, the Gibigiana is a table lamp that allows the user to direct a concentrated beam of light by adjusting a mobile mirror. The name references a playful optical phenomenon—*gibigiana* in the Milanese dialect—referring to light reflected from a shiny surface. Among the artefacts examined during the project, the Gibigiana was notable for the richness of its documentation, comprising a substantial number of preparatory drawings, photographic slides, and intermediate models preserved at the Fondazione Achille Castiglioni.

Given the nature of the materials, a tailored digitisation protocol was implemented. Large-format technical drawings (A0/A1), which did not require extremely high resolution, were digitised using digital photography. These were captured using a professional-grade digital camera, with a minimum resolution of 240 dpi, in a dedicated photographic setup at LabImmagine, Department of Design, Politecnico di Milano. The intermediate models housed at the Fondazione were also digitised through digital photography. The final model was documented via multi-angle photography to generate a 360° virtual representation.

In parallel, Sacchi's wooden prototype of the Gibigiana was digitised using laser-based 3D scanning technology at the Department of Design's Virtual Prototyping and Reverse Modelling Laboratory. The process involved acquiring high-density point clouds, which were then post-processed to generate a clean polygonal mesh. Textures were applied to the mesh to produce a photorealistic 3D model optimised for on-screen visualisation.

Small-format drawings, concept sketches, slides, and photographic transparencies were digitised using a flatbed scanner, with resolutions ranging from 300 to 600 dpi for drawings and up to 4800 × 2400 dpi for transparencies. All scans were saved in uncompressed TIFF format to ensure archival quality. The digitisation of these more delicate originals was conducted directly at the Fondazione Achille Castiglioni, thereby minimising the risk of damage due to transportation.

The systematic review, ordering, and digitisation of these materials enabled a detailed reconstruction of the design process underlying the Gibigiana—from its earliest conceptual sketches to the final product. This process, almost organic in its development, reveals the evolving form of the object as a continuous narrative authored by Castiglioni himself. The materials—especially the hand-drawn sketches—bear traces of intuition, experimentation, reconsideration, and even moments of distraction, thus offering a rare insight into the cognitive and creative dimensions of design practice.

Leveraging the SIRBeC platform, 180 metadata records were created for the Gibigiana, now accessible through the Lombardia Beni Culturali digital portal. These records offer a structured view of the lamp's development across successive phases, providing a valuable resource not only for scholars and design historians, but also for students who benefit from a pedagogical approach grounded in both critical reflection (mind-on) and practical engagement (hands-on).

Notably, several examples within the dataset illustrate Castiglioni's bricoleur sensibility (Lévi-Strauss 2003)—his capacity to appropriate existing objects or fragments in the creation of new, coherent forms. Some interventions were prompted by technical requirements, such as cost reduction or material optimisation; others display stylistic continuities with projects developed later in his career. In some instances, references to future design directions can be discerned in the margins of drawings—visual footnotes that foreshadow ideas yet to be realised (Fig. 2). Such findings exemplify the interpretive richness of archival design research, especially when supported by methodologically robust digitisation and cataloguing strategies.

5 Looking for Achille Castiglioni: Situated Storytelling for Design Heritage

Several months of immersive work within the Fondazione Achille Castiglioni made evident the extraordinary potential of the archival materials preserved there—not only those related to widely recognised canonical works, but also lesser-known projects that reveal the depth and diversity of the designer's intellectual and creative production. In alignment with the mission of the Fondazione Achille Castiglioni to disseminate the designer's legacy, the research team developed Looking for Achille Castiglioni (LfAC), a smartphone application designed to present Castiglioni's work situated in the urban fabric of central Milan (Spallazzo et al. 2011).

The application guides users through a curated urban itinerary comprising buildings, temporary exhibitions, installations, and interventions that, in many cases, no longer physically exist in their original form or location. By activating geolocated multimedia content, the app enables users to re-experience these lost or hidden works through an interpretive narrative. As the user approaches specific geographic coordinates—identified via the phone's GPS sensor—the application triggers a cluster of content tied to that site, including audio commentary, historical images,

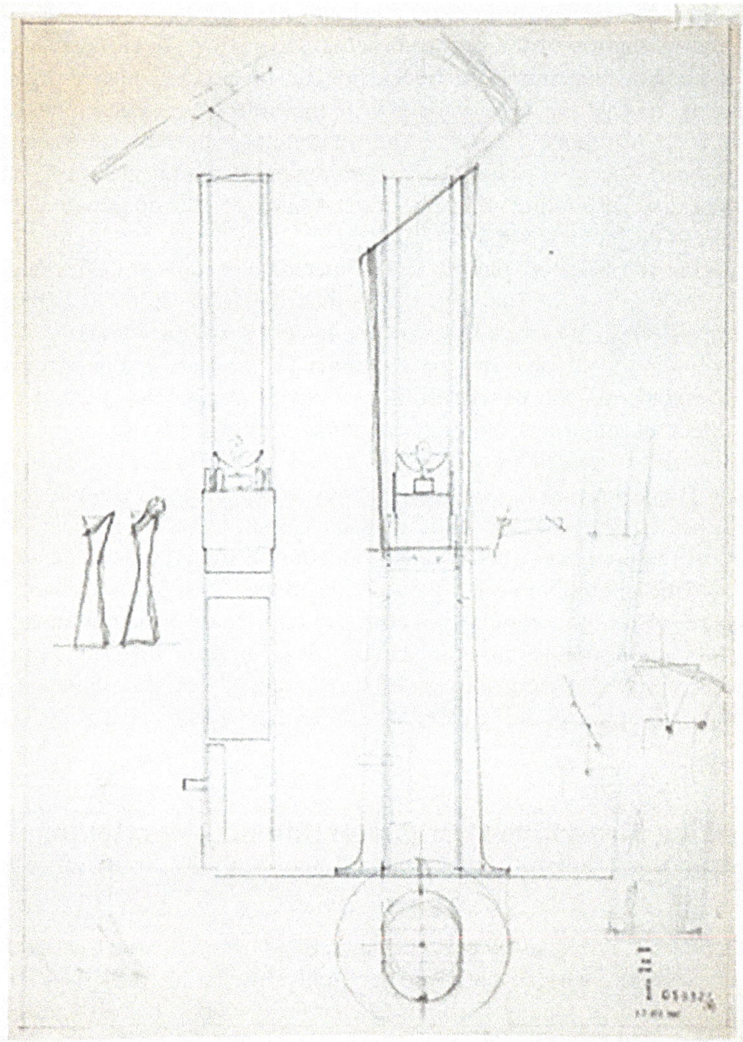

Fig. 2 Sketches and brainstorming on the Gibigiana lamp project. (Image courtesy of Fondazione Achille Castiglioni—SIRBEC)

sketches, videos, and project documentation. This geospatial model not only facilitates a self-guided tour but also collects data on user engagement and interaction patterns, providing feedback for ongoing evaluation of the project's communicative effectiveness.

The selection of Points of Interest (PoIs) was conducted in close collaboration with the curators of the Fondazione Achille Castiglioni. Criteria for inclusion included walking distance from the original studio and the availability of archival documentation to support a rich narrative experience. Fifteen PoIs were ultimately selected. Each PoI is equipped with a concise audio guide in both Italian and English, supplemented by visual content (e.g., photographs, drawings, design renderings), thus enabling the construction of an iconographic and interpretive narrative. The tour, designed to take approximately 1 h (excluding walking time between PoIs), is organised thematically—covering architecture, industrial design, and interiors—with each theme visually differentiated through distinct colour codes within the app interface.

For each PoI, a short video is launched with an audio narration describing the corresponding project. Simultaneously, related images and sketches are displayed on the user's device, enriching the interpretive framework with both verbal and visual layers. The experience is explicitly designed for an audience of informed users—design professionals, students, and enthusiasts—seeking in-depth knowledge and novel insights into Castiglioni's practice. Accordingly, the language of the audio tracks is calibrated to this audience, incorporating precise technical terminology and detailed contextual information (Fig. 3).

Structurally, the narrative of the application is composed of a series of micro-stories, each offering a window into Castiglioni's distinctive design approach. These short stories are non-linear, allowing users to construct their own thematic or spatial trajectories through the city based on personal interest and physical location. This format supports a hybrid form of cultural participation—both exploratory and reflective—anchored in place-based interaction with the city's architectural and cultural memory.

Fig. 3 LfAC Looking for Achille Castiglioni Mobile app

From a stakeholder perspective, the principal actors involved were the Fondazione Achille Castiglioni and its diverse public. The application offered a novel channel for the Fondazione to extend its engagement with visitors, complementing traditional archival and exhibition experiences. Importantly, the project facilitated the contextual relocation of archival content, situating it in the very environments where Castiglioni's works were originally conceived or installed. In doing so, it challenged the conventional museographic paradigm of displaced artefacts by proposing a situated form of digital storytelling.

The research team's role included not only technological and design development, but also the translation of complex and heterogeneous archival material into a coherent and compelling narrative format. The curators' expertise was essential in this process, ensuring the interpretive accuracy and cultural relevance of the selected content.

User feedback collected during testing phases revealed a strong interest in this form of augmented cultural experience, particularly among those who had previously visited the Studio Castiglioni. For these users, the app offered an expanded and contextualised understanding of the designer's work and methods. The project thus demonstrates the potential of mobile and location-based technologies to support the dissemination of design heritage, particularly through the reconstruction and narration of the creative process as it unfolds in drawings, models, and photographs. Such initiatives underscore how the legacy of iconic designers can be meaningfully reactivated through immersive and participatory forms of engagement grounded in archival research.

6 Conclusion

The valorisation of cultural heritage through archival materials enables a profound re-engagement with the *faber*'s ethos—his inventiveness, technical skill, procedural logic, and communicative vision. By digitising these materials, often large in format or inherently fragile, institutions can ensure both preservation and accessibility. The digital surrogates make visible what would otherwise remain hidden in storage or at risk of degradation, bringing renewed attention to artefacts that embody the cognitive and material traces of design practice.

Just as writers express their thinking through language, the *Maestri* of Italian industrial design—figures such as Achille Castiglioni and Giovanni Sacchi—communicated their creative intelligence through drawings, models, and prototypes. These artefacts are not merely documentation but repositories of a distinctive design culture, one that has shaped Italian identity on a global stage. Although the methodology presented here represents only one among many possible strategies for cultural heritage valorisation, it demonstrates the potential of digital tools to fulfil these aims when combined with thoughtful curatorial and narrative frameworks.

By re-focusing attention on the design process rather than solely on final products, the digital enhancement of archival material can enrich the interpretive experience of design exhibitions. More importantly, it offers a pathway to elevate processual knowledge as part of cultural heritage in its own right—bridging memory, creativity, and pedagogy. Given the relative affordability and replicability of such digitisation approaches, public and private institutions alike could more systematically promote the representation of the design process as a key heritage value.

In this light, digitisation is not merely a technical procedure but a means of reviving the spirit of the faber, allowing their knowledge and methods to circulate, inform, and inspire. This is a heritage that deserves to be not only preserved but rediscovered and reactivated. Nevertheless, challenges remain. These archival documents—technical drawings, prototypes, marginal sketches—are often difficult for non-expert audiences to interpret, which risks limiting the impact and reach of even the most comprehensive digitisation projects. Their full potential remains largely untapped, particularly in terms of public engagement and educational use.

As Martien De Vletter, Associate Director of the Canadian Centre for Architecture, insightfully observes: "Finding ideas in a collection based on records with or without visual support will no doubt continue to be a creative and intellectual process that requires inventive researchers that are able to give meaning to collection material and to make connections between materials that others overlook" (De Vletter 2020). In this perspective, the task of digitisation is only the beginning. It is the interpretive, connective, and narrative labour—undertaken by curators, researchers, and designers—that ultimately transforms archival material into cultural heritage that speaks to both contemporary and future audiences.

As these cases have shown, through careful selection, methodological rigour, and context-sensitive design, digitisation can offer new life to the creative legacies of the faber and shape the ways in which we understand, teach, and experience design history.

Competing Interests The authors have no conflicts of interest to declare that are relevant to the content of this chapter.

References

Angelaki G et al (2010) ATHENA: a mechanism for harvesting Europes's museum holdings into Europeana. Archives and Museum Informatics, Denver

Antonelli P, Guarnaccia S (2000) Achille Castiglioni. Corraini, Mantova

Auer P et al (1998) Normativa per l'acquisizione digitale delle immagini fotografiche. ICCD, Roma

Consalez L, Bertazzoni L (1998) Modelli e prospettive. Hoepli, Milan

De Vletter M (2020) CCA. Canadian Center for Architecture. In: Pavoni R, Riccini R (eds) Narrare con l'archivio. Fondazione Vico Magistretti, Milano

Degiarde E (2007) SIRBeC-Sistema Informativo Regionale dei Beni Culturali: metodologie e strumenti per la qualità del processo di catalogazione e valorizzazione delle informazioni prodotte. IReR, Milano

Greenfield A (2006) Everyware: the dawning age of ubiquitous computing. New Riders Publishing, Berkeley

Lévi-Strauss C (2003) Il pensiero selvaggio. Net, Milan

Maldonado T (1987) Questioni di similarità. Rassegna 32:57–61
Pavoni R, Riccini R (2020) Narrare con l'archivio. Fondazione Vico Magistretti, Milan
Polano S (2006) Achille Castiglioni. 1918-2002. Electa, Milan
Polato P (1991) Il modello nel design. La bottega di Giovanni Sacchi. Hoepli, Milan
Salvarani F (1992) La costruzione del modello architettonico. Di Baio Editore, Milan
Spallazzo D et al (2011) Walking, learning, enjoying. Mobile technology on the trail of design masterpieces. In: Dellepiane M et al (eds) Vast 2011. 12th international symposium on virtual reality, archaeology and cultural heritage. Eurographic Association
Weiser M (1991) The computer for the 21st century. Sci Am 265:94–105

Milano and the Everyday Informal: Mapping the Intangible Heritage of Italian Design

Ilaria Bollati and Luisa Collina

> *Who is the author? The anonymous inventor?*
> *Piergiacomo? Achille?*
> *The builder? The model maker?*
> *The author is not a name.*
> *The author is the invisible weave of stories.*
> *The author is a relationship.*
>
> *Laura Curino, Mani grandi, senza fine (2011–2012)*

Abstract Intangible cultural heritage, particularly oral traditions, is crucial for understanding Made in Italy, yet it risks fading away without proper documentation tools. In Italian design, ideas are born not only in studios but also in informal spaces like Bar Jamaica, Plastic, Caffè Craja, and Santa Tecla—historic Milanese venues from the postwar era where designers and creatives came together to exchange ideas and experiment, shaping the evolution of Italian design. Our research explores these hidden networks, mapping the memories and relationships that have contributed to the rise of Italian design. This project, part of the Changes program (PNRR, Politecnico di Milano—Department of Design) in Spoke 2-Creativity and Intangible Cultural Heritage, aims to map the intangible heritage of design and craftsmanship. Through digital technologies, these stories and connections can be captured, recorded, and shared, creating a repository of human memories that might otherwise be lost and offering new cultural experiences accessible not only to experts in the field.

Keywords Design · Oral memory · Intangible · Digital · Human heritage

I. Bollati (✉) · L. Collina
Politecnico di Milano, Milan, Italy
e-mail: ilaria.bollati@polimi.it

P. Cordera (ed.), *Unfolding Made in Italy (1948-1962)*, Research for Development, https://doi.org/10.1007/978-3-032-10118-1_10

1 Design Is Born Through Relationships

Italian design—particularly the Milanese tradition—has never been a solitary act. It is not a self-contained idea, but rather an unfolding narrative: a dense weave of encounters, dialogue, attentive listening, and constant negotiation. It is a living process that evolves, a form of knowledge shared among designers, artisans, model makers, educators, and clients.[1]

In her theatrical monologue *Mani grandi, senza fine* or Large, endless hands—performed in 2011 at Milan's Piccolo Teatro—Laura Curino portrays authorship in Milanese design as an invisible network of stories, powerfully asserting that "the author is not a name, it is a relationship." It is a collective dimension, shaped by collaboration and continuous exchange. Curino paints the city itself as a vibrant, ever-moving organism: "This city of grey stones moves, swaying gently in its smog, in its noise, in its northern lights... electric lines, railways, trams, telephones... a tangle of pipes, cables, towers, antennas, spider webs, and trolleybuses... a city rising, weaving relationships kept alive even at a distance by phones that begin to work again."

Milan, then, is not merely a physical place, but a living narrative made of searching hands—a social and urban fabric rich with opportunities for encounter, nurturing design, and allowing it to evolve daily.

Consider Ettore Sottsass and his *Scritto di notte*, or Written at Night (Sottsass 2010), a profoundly personal account of the years Milanese design took shape. He intertwines his story with a city and an entire era. With irony and irreverence, Sottsass reminds us that every project is a mosaic of memories, lives, and influences: a story constantly in motion, never conclusive. Aligned with this vision, two additional theatrical experiences illuminate invisible lines and connections rooted in human and creative relationships.

In 1998, American director and playwright Robert Wilson created *70 Angels on the Façade*, a performance inspired by an idea from architect and designer Gio Ponti to mark the 70th anniversary of *Domus*. A continuous 7-day theatrical event blended natural and supernatural time, transforming the stage into a window on the world—where design becomes a lived experience and an ongoing conversation. More recently, in 2025, the installation *Staging Modernity* by Formafantasma, presented at the Teatro Lirico Giorgio Gaber in Milan, reinterpreted the iconic 1929 Salon d'automne floorplan. By fragmenting and dispersing the symbols of modernity, the installation offered a new lens through which to read them. The space was populated with animal replicas, aiming to dissolve the boundaries between human and non-human, modern and wild.

In these examples, design becomes porous. It allows itself to be traversed, absorbed, and transformed. There is no fixed boundary between natural and artificial, human and non-human: everything overlaps, blends, and is redefined. The project becomes an open space, inhabited by multiple relationships and

[1] On this topic in its historical context, see Cordera's essay in this volume.

presences—unseen yet deeply felt. These are relationships shaped by informal exchanges, fleeting conversations, and encounters that defy conventional structure but sustain, day by day, the most vital spirit of design thinking. It is this human fabric—fragile yet essential—that safeguards a valuable part of design's intangible heritage. And because it is so closely bound to the people who inhabit it, this heritage is in danger of disappearing unless it is actively and thoughtfully nurtured.

2 Informality as a Site of Knowledge

As Ettore Sottsass recounts in a memory that acts as vivid testimony:

> In the evening I'd go to the Jamaica or to the Sorelle Pirovini [Pirovini Sisters] [...] I, too, would go to the Pirovini to feed myself with pasta and beans without having to face the lady from Lugano. I learned about the possibility of credit much later, so I almost always paid for my pasta and beans. And if I had no money, I'd go to the Pirovini just to meet people. In my life, that big room at the Pirovini was very important, because I met many young people just beginning to live, and others – no longer so young – who had already clashed with a complicated and dangerous existence during the war. At the Pirovini, I met Vittorini, Alfonso Gatto, Dova, Crippa, Sergio D'Angelo, Capogrossi, Manzoni, Dorazio, Pontecorvo, Morlotti, Cassinari, Munari, Radice, Soldati, Fortini, Kodra, Dorfles, Max Huber, Giansiro Ferrata, and others – all agitated, all in a hurry, yet none of them were on track to become university professors, and hardly anyone had yet been published in newspapers, except perhaps Calvino, published by Vittorini in Il Politecnico. No one had finished their book, no one had written their poems, no one had yet entered the museums… (Sottsass 2010, pp. 150–151).

In these informal places, woven through with sociality and serendipitous encounters, a kind of knowledge was nurtured—not born in laboratories or design studios, but around café tables, shared kitchens, and in the living rooms of friends.[2] At its core, this research project by the Politecnico di Milano and Università degli Studi di Milano, funded by the PNRR as part of the CHANGES initiative—Cultural Heritage Active Innovation for Next-Gen Sustainable Society.[3] The project explores the city through its submerged geographies: social spaces where, from the postwar years to the present, thoughts, affections, and design practices have intertwined. In this relational perspective, design happens first and foremost between people—before it ever manifests in objects or spaces.

[2] For the display of Italian design in hotels, see the essay by Scullica and Elgani in this volume.

[3] Credits of the team involved in the conceptual, content-related, and organizational development of the project *Milano In-Formale*, without yet detailing the authorship or production of individual outputs. CHANGES Spoke 2 Scientific Director: Alberto Bentoglio (Università degli Studi di Milano), Project Scientific Responsibility: Luisa Collina (Politecnico di Milano), Scientific supervision: Giampiero Bosoni, Luisa Collina, Chiara Lecce (Politecnico di Milano) Editorial research and curatorship: Ilaria Bollati, Marta Elisa Cecchi (Politecnico di Milano), Research support: Alice Biancardi, Chiara Di Lodovico (Politecnico di Milano), Experience design and production coordination: Ilaria Bollati (Politecnico di Milano), Illustrations: Francesca Gastone (Atelier Fyumi).

Cafés like Jamaica, Craja, or Bar Basso, clubs like Plastic, and even studio-kitchens, backrooms, street corners, and shared apartments all form environments where words flow freely, ideas cross-pollinate, and insights settle far from the spotlight. These spaces do not demand immediate productivity but allow something new to emerge spontaneously. In these marginal moments of everyday life, Milanese design has often found its most profound and unexpected forms of innovation. As artist Enrico Baj recalled, "D'Angelo, Joe Colombo, and I did everything: we founded movements, issued manifestos, transformed and ran dance halls to the sound of boogie-woogie. We painted, played music, danced until morning—like in the Saint-Germain of Juliette Gréco and Boris Vian. Cellars and nightclubs like Arethusa, Santa Tecla, Mocambo, Taverna Mexico, the Shanghai with its blind pianist—Milan by night was all ours" (Baj 1983, pp. 176–177). These words evoke a city charged with intense cultural energy, where the boundaries between art, life, and design blur and significant creative practices are forged in the margins of the everyday (Fig. 1).

Within this context, leisure time becomes a precursor to design: seemingly unproductive time that is nonetheless fertile, generative, and transformative. It is no

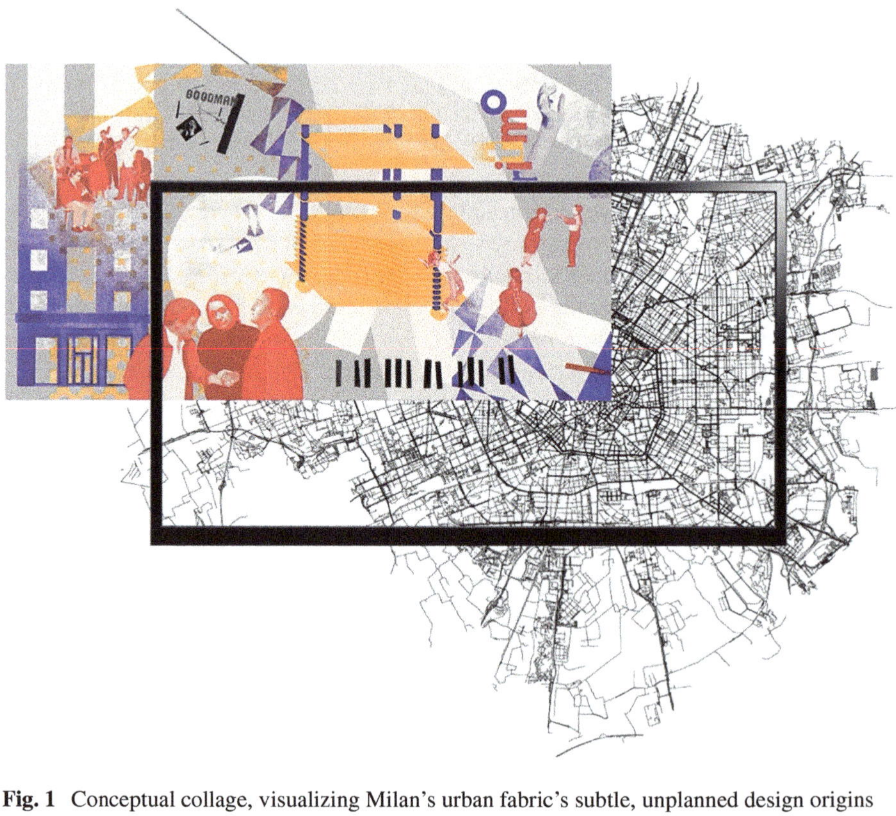

Fig. 1 Conceptual collage, visualizing Milan's urban fabric's subtle, unplanned design origins

coincidence that *Tempo libero* or Leisure was chosen as the 13th Milan Triennale theme in 1964. Leisure is the space of missteps, of uncertainty, of idle chatter that suddenly turns into an idea—echoing Massimo Recalcati's claim that "the misstep [becomes] the condition for research" (Recalcati 2014). Like any creative figure, the designer may be seen as a modern *flâneur*—an urban wanderer with a curious and attentive gaze, attuned to the fleeting and meaningful details of city life. Just as the *flâneur* sits at a café to watch the city flow by, the creative mind feeds on these slow, contemplative experiences, transforming them into new visions.

Beba Restelli, assistant to Bruno Munari, writes in *Giocare con i fili*, or Playing with Threads (Restelli 2023), that in cities like Vienna, Paris, or Lisbon, tables still exist that are dedicated to the most devoted *flâneurs*, and there are even statues that honour this wandering gaze. Milan is no exception: it, too, has spaces that have embraced such creative contemplation—places where time stretches and idleness becomes productive. Bruno Munari once described Bar Craja as "the most precise and clean place to meet friends"—an image that might seem outdated today but retains a rare clarity of thought, relationships, and choices. Designed by architect Luciano Baldessari,[4] along with Figini and Pollini, in 1930, Craja was "a small planetary station from which to depart for enlightened Europe" (Cimoli 2024): an open place where culture was lived as a political experience, deeply embedded in everyday life.

This exploration does not map the canonical sites of design but instead traces the escape points where design happened without planning, where the encounter preceded the idea. Within these fertile voids, suspended between attentiveness and leisure, new futures are still being cultivated today. These are not spaces that declare themselves productive, but they act in depth: sites of suspension and possibility, where the seeds of design nestle within ordinary gestures, purposeless conversations, and relationships that generate unprecedented ways of thinking. Here, design does not declare itself—it happens, quietly, in the folds of daily life.

Anna Maria Cimoli urges us to look again at Craja not as a nostalgic memory, but as a vital model: "Almost a century later, it seems important to study that model not for a nostalgic 'how we were', but to understand and transmit what still holds generative power—at a time of deep solitude and latent tensions that the very form of the city, on different scales, conveys to those who can read them" (Cimoli 2024, p. 12). From this arises the urgency to narrate informal places not as relics of the past, but as living sources of knowledge and intuition: suspended spaces that continue to offer essential potential to nourish innovation and vitality in Milanese design—transforming informality into a design resource. The echo of Munari reminds us that the past is not superficial or romantic nostalgia—it is a precious reservoir of knowledge, from which we can draw lessons for both the present and the future (Fig. 2).

[4] For references to the Baldessari archive at the CASVA, see Feraboli's essay in this volume.

Fig. 2 Initial visualization tests of possible key sites explored during the research. (Images by Atelier Fyumi). We explore key visual language elements to align the illustrations with all project outputs

3 Knowledge That Resists Archiving

And yet, this intangible heritage and oral knowledge—precisely because it does not immediately take the form of an object, a space, or a document—remains fragile, precarious, and exposed to oblivion. Glances nourish the relational heritage of design, shared emotions, and conversations that unfold at the project's margins yet constitute its deepest structure. It is a kind of knowledge that resists archiving and must instead be listened to, interpreted, and accompanied. As Giuseppe Panza di Biumo (2008) noted, "We know all that has reached us because we can see it, but there is a universe of things that never reached us and were equally fundamental in shaping what we know."

It is precisely this shadow zone—composed of ephemeral and invisible knowledge—that today challenges how we document and transmit immaterial forms of knowing. Oral memory is no less authoritative; it is simply more situated, vulnerable, and human. For this reason, its preservation cannot be automated or neutral: it is a labour of relationship, trust, and empathy. Lidia Decandia, in her study *Portatori di storie*, or Bearers of Stories (Decandia 2011), describes those who safeguard and transmit local knowledge through oral storytelling. This work is not merely about collecting testimonies but about recognizing those presences that shape a relational epistemology, rooted in lived experience and affective bonds.

A telling example of the collective and intentionally "imprecise" nature of such knowledge appears in the disclaimer of the exhibition dedicated to Elio Fiorucci at the Milan Triennale (2024–2025), where curators explain that "given Fiorucci's typical working method—based on a collective creative language—it was often difficult to identify definitive information about individual works or to attribute a specific author to each project." This underscores how the informality of this heritage is nourished by collectivity, word of mouth, the said and unsaid, and how the aura made up of legends, memories, and overlapping recollections is an essential part of its identity.

Similarly, works like *Louvre, mon amour* by Pierre Schneider (2023) and *Caffè Paradiso* by Massimiliano Gioni (2024) place conversation and oral heritage at the centre. Both books revolve around emblematic places of leisure—the Louvre Museum and Caffè Paradiso at the Venice Biennale—and, through dialogues, pauses, glances, and silences, they chart intimate maps of thoughts, emotions, and creative processes. Gioni recounts how Caffè Paradiso served for decades as a meeting point for curators and artists—a vital place of exchange for navigating the cultural and organizational challenges of the Biennale—highlighting the importance of the human dimension behind major cultural events. And yet, neither books nor traditional archives can fully capture the human voice in all its richness: tone, modulation, cadence—elements that profoundly affect the effectiveness, credibility, and memorability of storytelling. In her article *È il momento della voce*, or It's the Time of the Voice, Barbara Meneghel (2021) emphasizes how voice is a medium of oral art, capable of giving shape to an experiential and subjective dimension of knowledge—one that is difficult to convey through written or documentary forms. The voice carries emotional weight and intimacy, becoming a connection and cultural transmission vehicle.

In this way, oral, verbal, and human content reveals a fundamental dimension of cultural and design knowledge—one that transcends what is documented or visible. Its collection requires careful practices sensitive to the contexts in which stories are born and shared. An oral transmission can be seen as a performative act that constantly regenerates knowledge, enabling collective and individual memory circulation and preserving a living heritage in continuous transformation.

Only through empathetic listening, active participation, and attentive mediation can verbal heritage be restored to its full dignity—not as a mere repository of information, but as a living and dynamic source of knowledge, capable of deeply enriching cultural and design practices. Even digital tools can play a key role, provided they are conceived not as neutral instruments but as sensitive infrastructures—capable of amplifying the quality of listening, detecting nuance, and revealing the depth of relationships. When integrated with care, they enable us to collect, preserve, and share the subtleties of oral communication—an intangible heritage that must not be stripped of its nature—thus making accessible a fragile, sometimes intimate, yet essential form of knowledge.

4 Podcasts and Sound Books: Giving Voice to Relationships

In French, *public* can mean many things. It refers to what is shared, a collective that listens or participates, but also the *bruit public*—that gossip that circulates through environments, communities, and social spaces. An evocative image that recalls orality, sidelong glances, and stories passed along outside official channels. This idea has accompanied us from the beginning of the project, which is dedicated to those who live and build culture beyond canonical categories. As Bätschmann (2024) writes, "like other publics, the art public is neither homogeneous nor autonomous."

We start from this: the public is not a monolithic block but a mutable constellation of different knowledges and intensities. The curious, the enthusiast, the expert, the designer, the student, those who work in the field, and those who do not. This project was born precisely for them, moving between research, accessibility, and cultural narration. We represented this pluralism with three autonomous and interconnected forms: an online map, a podcast series, and a sound book. Three ways to approach the content, three experiences that do not overlap but rather complement and enrich one another. Each is designed to activate different modes of engagement: visual, auditory, intellectual, or emotional.

The illustrated digital map (Fig. 3) is the visual heart of the project: an online cultural experience that organizes and makes accessible, chronologically and geographically, the relationships between places, people, objects, and design projects. It is not a static atlas but a journey through time that shows the evolution and intertwining of these worlds. The original and specially created illustrations do not explain but evoke. They do not describe but suggest.

The podcast, available on Spotify, consists of five episodes that follow a narrative order different from the chronological order of the map. Here, the sequence is built around curiosity and the common imagination of the public: if the map begins with Bar Craja as the origin, the podcast instead starts from iconic places such as Jamaica, Plastic, or Bar Basso. This choice aims to hook the listener with an engaging and accessible story, based on themes and spaces that resonate with the contemporary

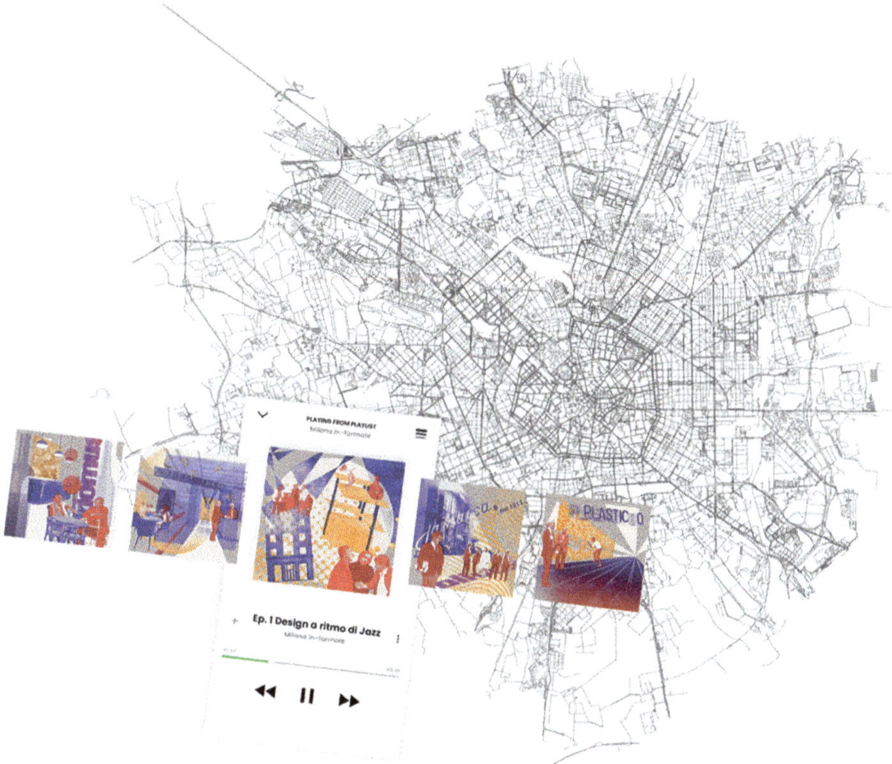

Fig. 3 Conceptual collages showing the interplay of anecdote, oral storytelling, podcast, and mapping

audience. Studies such as the Ipsos Digital Audio Survey (2023) confirm the effectiveness of micro-series of 4–5 episodes and podcasts of limited duration, which keep attention high and stimulate continuous listening.

The voices heard are multiple: those who built those places, passed through, inhabited, or observed them. It is a multilayered choral story that creates familiarity with what often remains invisible or informal. Listening thus becomes an immersive experience, capable of breaking down the barriers between knowledge and curiosity, authority and participation. Finally, the sound book represents a third, more experimental and transversal experience. It follows neither the chronological order of the map nor the podcast's narrative. Still, it tells the deeper themes and hidden threads that connect them, offering a reflection that traverses and underlies the other two levels. An experimental editorial object that combines the physicality of the pages to be flipped, the object's design, and the scientific book's concreteness with the immaterial lightness of the podcast, proposing a different and complementary access to the same universe.

In this system, each element is autonomous but connected. The map leads to the podcast, the podcast refers to the book, and the book reconstructs paths that return to the map. No fixed path exists: you can start anywhere, like in a constellation. Everything holds together without rigidity. An invitation to look, listen, explore—without hierarchy. And, above all, to participate. Because once again, the public is not an entity to be defined but a multitude of perspectives to welcome and stimulate.

5 Traces for New Worlds

Interpretation aims to enable visitors to experience genuine intellectual and emotional gratification. Carrada emphasizes:

> We may be the best at studying, protecting, and restoring, but we still have not learned to interpret – that is, to connect objects and places with visiting people. To offer them new knowledge and curiosity. To awaken emotions, imagination, and inspiration. To make them want to see, know, and enjoy more. […] To interpret means to make people understand and enjoy together, because there can be no understanding without emotion (Carrada 2025, pp. 3–4).

From these insights, this project takes shape: not a completed work, but a living, porous device that generates more questions than answers. A cultural gesture that starts from an idea now widely shared even in academic circles: knowledge is never definitive; it is always partial, situated, and evolving (Haraway 1988). This perspective resonates with Recalcati's vision when he writes that knowledge is "pierced" and, by nature, incomplete. It is precisely this structural gap that ignites the desire to understand. A desire nourished by freedom, slow rhythms, and places outside the frame. Informal spaces that, precisely because they are marginal, become surprising generators of meaning.

The mapping we proposed does not follow an encyclopaedic order. Still, it takes the form of a dérive, as Guy Debord (2001) conceived it: relying on intuition, details, and memories. It speaks to specialists and invites everyone to observe the city differently—with the gaze of a designer lingering in a bar, a student choosing the same bench every day, a researcher who listens rather than explains.

The podcast and the sound book also move along this trajectory. The former follows emotional threads, fragments of lived experience, and narrative suggestions. The latter mixes voice and written word, allowing an experience of listening and reading that stimulates imagination more than analysis. As Sterne (2003) noted, sound is often underestimated in cognitive processes, but in reality, it enhances our capacity to feel and interpret.

This investigation does not aim to explain everything. On the contrary, it starts from the awareness, as Recalcati also writes, that "Knowledge is not only about accumulating information, but about opening to desire—an openness that allows other worlds to emerge beyond the ones we already know" (Recalcati 2014).

In this light, emptiness—as a blank canvas—is not absence but potential. It is a space already inhabited by traces, memories, and expectations. Like every blank sheet in the drawer, it is dense with what came before. The map gathers and reactivates these fragments—not to archive but to make them speak again. The Turkish writer Orhan Pamuk says that "the journey toward the unknown is what gives meaning to the narrative" (Pamuk 2014). This impulse guides the project: to create a space open to the unsaid, the imperfect, the fragile. An interpretation as a form of restitution, a restitution that does not close but relaunches.

As researchers, teachers, designers, and citizens, our role is not to guard a definitive truth but to create the conditions for desire to emerge. Because, as Paolo Rumiz says, "there are geographies that exist only if someone tells their story" (Rumiz 2023).

This mapping traces just one of these: made of pauses, detours, intuitions. A narrative that moves between memory and possibility, between knowledge and listening. And if, as Recalcati continues to ask, "is a life not played entirely in the contingency of encounter?" (Recalcati 2014), then perhaps our responsibility is this: to make space for that encounter. With the city. With the other. With a voice.

Competing Interests The authors have no conflicts of interest to declare that are relevant to the content of this chapter

References

Baj E (1983) Automitobiografia. Rizzoli, Milan
Bätschmann O (2024) Il pubblico dell'arte: Una breve storia. Johan & Levi, Milan
Carrada G (2025) Perché non parli? Come raccontare il patrimonio culturale. Johan & Levi, Milan
Cimoli AC (2024) Il tempo del Craja: Biografia di un caffè. Nomos Edizioni, Busto Arsizio
Debord G (2001) Teoria della deriva. In: Careri L (ed) I Situazionisti e la deriva urbana. DeriveApprodi, Roma. (Opera originale pubblicata nel 1956)
Decandia L (2011) Sensitive city: La città dei portatori di storie. Intrecciare arte, memoria e nuove tecnologie per costruire nuovi modi di narrare la città. In: Abitare l'Italia: Territori, economie, diseguaglianze. XIV Conference SIU, 24–26 March 2011. https://www.planum.net. Accessed 20 Aug 2025
Gioni M (2024) Caffè Paradiso: La Biennale di Venezia raccontata dalle sue direttrici e dai suoi direttori. Johan & Levi, Milan
Haraway D (1988) Situated knowledges: the science question in feminism and the privilege of partial perspective. Fem Stud 14(3):575–599. https://doi.org/10.2307/3178066
Ipsos Digital Audio Survey (2023) Ascolto e fruizione dei podcast: Ipsos presenta i risultati della Digital Audio Survey 2023. https://www.ipsos.com/it-it/ascolto-fruizione-podcast-ipsos-digital-audio-survey. Accessed 20 Aug 2025
Meneghel B (2021) È il momento della voce. Rivista Studio, Milan. https://www.rivistastudio.com/arte-orale-voce/. Accessed 20 Aug 2025
Pamuk O (2014) A modest manifesto for museums. The Museum of Innocence. https://www.masumiyetmuzesi.org/en/mani-festo. Accessed 20 Aug 2025
Panza di Biumo G (2008) Light art: Considerazioni per un archivio dedicato alla luce. Maggioli, Santarcangelo di Romagna
Recalcati M (2014) L'ora di lezione. Einaudi, Turin
Restelli B (2023) Giocare con i fili. Franco Angeli, Milan
Rumiz P (2023) Una voce dal profondo. Giangiacomo Feltrinelli, Milan
Schneider P (2023) Louvre, mon amour: Undici grandi artisti in visita al museo più famoso del mondo. Johan & Levi, Milan
Sottsass E (2010) Scritto di notte. Adelphi, Milan
Sterne J (2003) The audible past: cultural origins of sound reproduction. Duke University Press, Durham

www.ingramcontent.com/pod-product-compliance
Ingram Content Group UK Ltd.
Pitfield, Milton Keynes, MK11 3LW, UK
UKHW020934180926
475692UK00003B/38

* 9 7 8 3 0 3 2 1 0 1 1 7 4 *